NATEF Standards Job Sheets

Brakes

(Test A5)

Jack Erjavec

THOMSON
DELMAR LEARNING

Australia Canada Mexico Singapore Spain United Kingdom United States

THOMSON

DELMAR LEARNING

NATEF Standards Job Sheets

Brakes
(Test A5)

Jack Erjavec

Business Unit Director:
Alar Elken

Executive Editor:
Sandy Clark

Acquisitions Editor:
Sanjeev Rao

Team Assistant:
Matthew Seeley

Executive Marketing Manager:
Maura Theriault

Marketing Coordinator:
Brian McGrath

Channel Manager:
Fair Huntoon

Executive Production Manager:
Mary Ellen Black

Production Manager:
Larry Main

Production Editor:
Betsy Hough

Cover Design:
Michael Egan

Printed in United States of America
6 XXXX 06 05 04

For more information contact Delmar Learning, Executive Woods, Maxwell Drive, Clifton Park, NY 12065-2919. Or find us on the World Wide Web at http://www.delmarlearning.com.

ISBN: 0-7668-6371-9

CONTENTS

PREFACE

With every passing day it gets harder to learn all that it takes to become a competent automotive technician. Technological advancements have allowed automobile manufacturers to build safer, more reliable, and more efficient vehicles. This is great for consumers, but along with each advancement comes a need for more knowledge.

Fortunately, students don't need to know it all. In fact, no one person knows everything about everything in an automobile. Although they don't know everything, good technicians do have a solid base of knowledge and skills. The purpose of this book is giving students a chance to develop all the skills and gain the knowledge of a competent technician. It is also the purpose of the guidelines established by the National Automotive Technicians Education Foundation (NATEF).

At the expense of much time and the work of many minds, NATEF has assembled a list of basic tasks for each of its certification areas. These tasks identify the basic skills and knowledge levels of competent technicians. The tasks also identify what is required for a student to start a successful career as a technician.

Most of what this book contains is job sheets. These job sheets relate to the tasks specified by NATEF. The main considerations in the creation of these job sheets were student learning and program certification by NATEF. Students are guided through standard industry-accepted procedures. While they are progressing, students are asked to report their findings and offer their thoughts on the steps they have just completed. The questions asked of the students are thought provoking and require students to apply what they know to what they observe.

The job sheets were designed to be generic; that is, whenever possible, the tasks can be performed on any vehicle from any manufacturer. Completion of the sheets does not require the use of specific brands of tools and equipment; rather, students use what is available. In addition, the job sheets can be used as a supplement to any good textbook.

Words to the Instructor I suggest you grade these job sheets on completion and reasoning. Make sure the students answer all questions, and then look at the reasons to see if the task was actually completed and to get a feel for their understanding of the topic. It'll be easy for students to copy others' measurements and findings, but each student should have his or her own base of understanding, and that will be reflected in the explanations given.

Words to the Student While completing the job sheets, you have a chance to develop the skills you need to be successful. When asked for your thoughts or opinions, think about what you observed. Think about what could have caused those results or conditions. You are not being asked to give accurate explanations for everything you do or everything you observe; you are only asked to think. Thinking leads to understanding. Good technicians are good because they have a basic understanding of what they are doing and of what they are doing it to.

NOTICE: SOME PARTS OF THIS COPY ARE DIFFERENT THAN THE PREVIOUS PRINTING OF THIS BOOK. THE CONTENTS HAVE BEEN UPDATED IN RESPONSE TO THE RECENT CHANGES MADE BY NATEF. ***SEE PAGE 163 FOR DETAILS.***

BRAKE SYSTEMS

To prepare you to learn what you should learn from completing the job sheets, some basics must be covered. This discussion begins with an overview of brake systems (Figure 1). Emphasis is placed on what they do and how they work. This includes the major components and designs of brake systems and their role in the efficient operation of brake systems of all designs.

Preparing to do something on an automobile would not be complete if certain safety issues were not addressed. This discussion covers those things you should and should not do while working on brake systems. Included are proper ways to deal with hazardous and toxic materials.

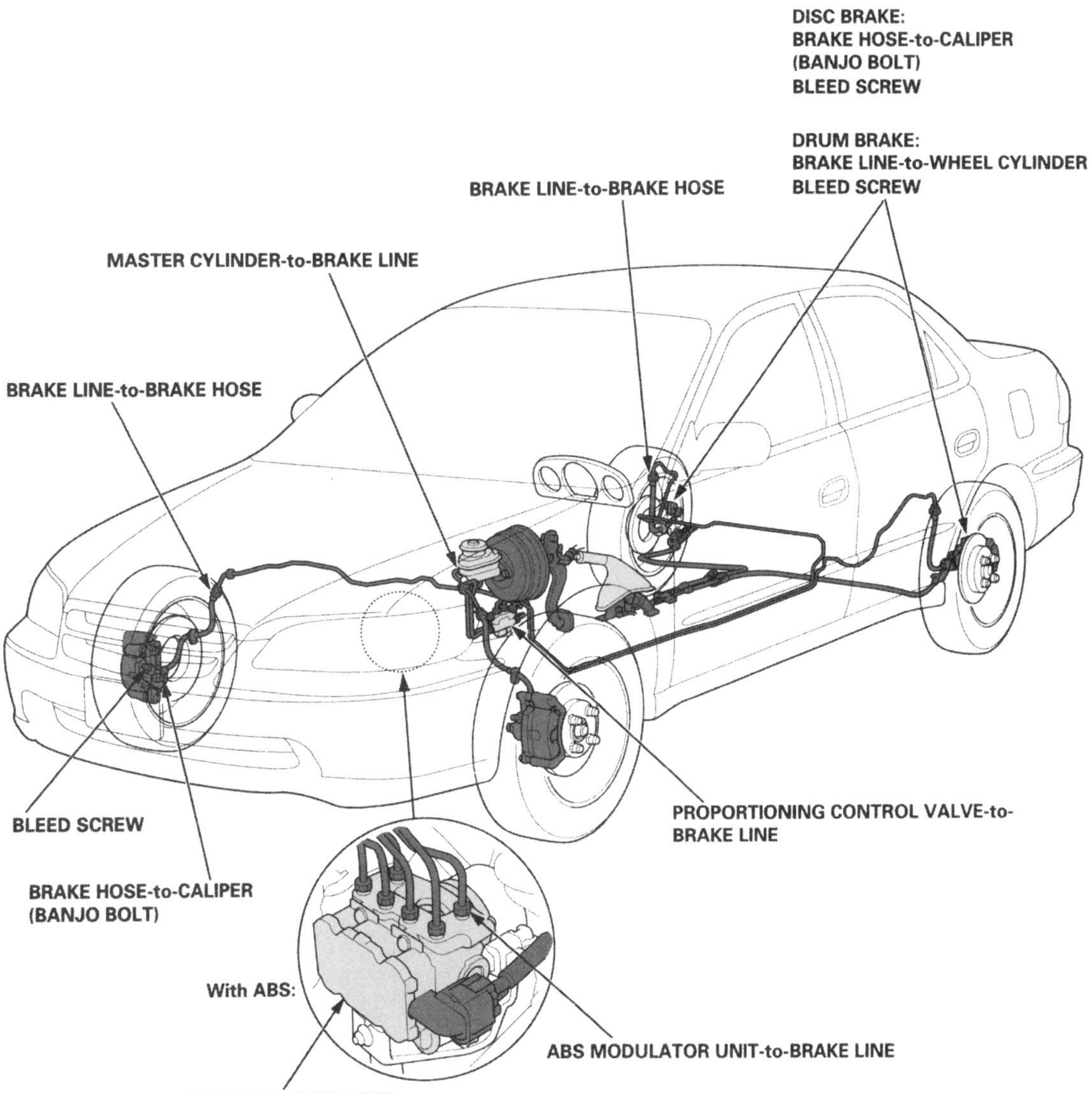

Figure 1 A typical brake system. *Courtesy of American Honda Motor Co.*

NATEF's task list for Brake Systems certification is also given with definitions of some of the terms used to describe the tasks. This list gives you a good look at what the experts say you need to know before you can be considered competent to work on brake systems.

Following the task list are descriptions of the various tools and types of equipment you need to be familiar with. These are the tools you will use to complete the job sheets. They are also the tools NATEF has identified as being necessary for servicing brake systems.

Following the tool discussion is a cross-reference guide that shows which NATEF tasks are related to specific job sheets. In most cases there are single job sheets for each task. Some tasks are part of a procedure and when that occurs, one job sheet may cover two or more tasks. The remainder of the book contains the job sheets.

BASIC BRAKE SYSTEM THEORY

As the brakes on a moving automobile are applied, rough-textured pads or shoes are pressed against rotating parts of the vehicle—either rotors (discs) or drums. The kinetic energy, or momentum, of the vehicle is then converted into heat energy by the kinetic friction of rubbing surfaces and the car or truck slows down.

When the vehicle comes to a stop, it is held in place by static friction. The friction between the surfaces of the brakes, as well as the friction between the tires and the road, resist any movement.

There are four basic factors that determine the braking power of a system. The first three factors govern the generation of friction: pressure, coefficient of friction, and frictional contact surface. The fourth factor is a result of friction—heat, or more precisely heat dissipation.

The amount of friction generated between moving surfaces contacting one another depends in part on the pressure exerted on the surfaces. In automobiles, hydraulic systems provide application pressure. The amount of friction between two surfaces is expressed as a coefficient of friction. The coefficient of friction is determined by dividing the force required to pull an object across a surface by the weight of the object. The amount of frictional surface, or area, that makes contact also determines braking effectiveness. Simply put, bigger brakes stop a car more quickly than smaller brakes used on the same car. Similarly, brakes on all four wheels slow or stop a moving vehicle faster than brakes on only two wheels, assuming the vehicles are equal in size.

The tremendous heat created by the rubbing brake surfaces must be conducted away from the pad and rotor (or shoe and drum) and be absorbed by the air. The weight and potential speed of a vehicle determines the size of the braking mechanism and the friction surface area of the pad or shoe. Brakes that do not effectively dissipate heat experience brake fade during hard, continuous braking. The linings of the pad and shoe become glazed, and the rotor and drum become hardened. Therefore, the coefficient of friction is reduced and excessive foot pressure must be applied to the brake pedal to produce the desired braking effect.

Brake Lining Friction Materials

The friction materials used on brake pads and shoes are called brake linings. Brake linings are either riveted or bonded to the backing of the pad or shoe. Some newer brake pads are integrally molded. These are identifiable by looking at the backing of the pad. Integrally molded pad assemblies will have holes that are partially or totally filled with the lining material.

Fully metallic linings of sintered iron have great fade resistance. However, they require very high pedal pressure and tend to quickly wear out drums and rotors. Semimetallic linings are made of iron fibers molded with an adhesive matrix. Semimetallic material offers excellent fade resistance with good frictional characteristics.

Principles of Hydraulic Brake Systems

The hydraulic system uses brake fluid to transfer pressure from the brake pedal to the pads or shoes. This transfer of pressure is reliable and consistent because liquids are not compressible. That is, pressure applied to a liquid in a closed system is transmitted by that liquid equally to every other part of that system.

Of course, the hydraulic system does not stop the car all by itself. In fact, it really just transmits the action of the driver's foot on the brake pedal out to the wheels. In the wheels, sets of friction pads are forced against rotors or drums to slow their turning and bring the car to a stop. Mechanical force (the driver stepping on the brake pedal) is changed into hydraulic pressure, which is changed back into mechanical force (brake shoes and disc pads contacting the drums and rotors).

Dual Braking Systems

Since 1967, federal law has required that all cars be equipped with two separate brake systems. If one circuit fails, the other provides enough braking power to safely stop the car. The dual system differs from the single system by employing a tandem master cylinder, which is essentially two master cylinders formed by installing two separate pistons and fluid reservoirs into one cylinder bore. Each piston applies hydraulic pressure to two wheels.

In early dual systems, the hydraulic circuits were separated into front and rear sections. Both front wheels were on one hydraulic circuit and both rear wheels on another. If a failure occurred in one system, the other system was still available to stop the vehicle. However, the front brakes do approximately 70% of the braking work. A failure in the front brake system would only leave 20 to 40% braking power. This problem was somewhat reduced with the development of diagonally split systems.

The diagonally split system operates on the same principles as the front and rear split system. It uses primary and secondary master cylinders that move simultaneously to exert hydraulic pressure on their respective systems. The hydraulic brake lines on this system, however, have been diagonally split front to rear (left front to right rear and right front to left rear). The circuit split can occur within the master cylinder or externally at a proportioning valve or pressure differential switch. In the event of a system failure, the remaining good system would do all the braking on one front wheel and the opposite rear wheel, thus maintaining 50% of the total braking force.

Brake Fluid

There are three basic types or classifications of hydraulic brake fluids. DOT 3 is a conventional brake fluid and is generally recommended for most ABS systems and some power brake setups. DOT 4 is also a conventional brake fluid and is the most commonly used brake fluid for conventional brake systems. DOT 5 is a unique silicone-based brake fluid and currently is not widely recommended by manufacturers. Only use the recommended and approved brake fluid in a brake system.

Brake Pedals

The brake pedal is where the brake's hydraulic system gets its start. When the brake pedal is depressed, force is applied to the master cylinder. On a basic hydraulic brake system (where there is no power assist), the force applied is transmitted mechanically. As the pedal pivots, it becomes a lever and the force applied to it is multiplied mechanically. The force that the pushrod applies to the master cylinder piston is, therefore, much greater than the force applied to the brake pedal.

Master Cylinders

A master cylinder converts the mechanical force of the driver's foot to hydraulic pressure. It has a bore that contains an assembly of two pistons and return spring. One piston pressurizes one-half of the brake system and another takes care of the other half. Although master cylinders differ in terms of the size of the pistons, reservoir design, and integrated hydraulic components, the operation of all master cylinders is basically the same.

The brake pedal, connected to the master cylinder's piston assembly by a pushrod, controls the movement of the pistons. As the pedal is depressed, the piston assembly moves. The pistons exert pressure on the fluid, which flows out under pressure, through the fluid outlet ports into the rest of the hydraulic system. When the brake pedal is released, the return spring in the cylinder forces the piston assembly back to its original position.

Each piston has a primary and a secondary cup to keep the cylinders sealed and filled with fluid. The primary cup compresses the fluid when the pedal is depressed. It keeps the brake fluid ahead of the pistons when they are put under pressure. The movement of the piston and cup shut off the fluid supply from the reservoir and create a sealed system from the primary cup forward. To prevent the fluid from leaking out of the cylinder when the primary cup has moved forward, each piston has a secondary cup. The secondary cup seals non-pressurized fluid.

When the brake pedal is not depressed, fluid enters the cylinder through vent (intake) ports and compensating or replenishing ports that connect the reservoir to the piston chamber. Each piston has a return spring in front of it. This holds the primary piston cup slightly behind the compensating or replenishing port for the reservoir, allowing gravity to keep the cylinder filled with fluid. In addition to keeping the replenishing ports uncovered, the return springs also help to return the brake pedal when the force has been removed from it. As the brakes are applied, the stiffer primary piston spring pushes the secondary piston and spring slightly. Then the cup at the front end of the secondary piston passes and closes off the primary replenishing port on the secondary side of the master cylinder.

When all of the brakes are fully released, any excess fluid is returned to the reservoir through the compensating port to relieve pressure in the system.

The two pistons in the master cylinder are not rigidly connected. Each piston has a return spring with the primary piston spring between the two pistons. Stepping on the brake pedal moves a pushrod, causing the first or primary piston to move forward. The fluid ahead of it cannot be compressed, so the secondary piston moves. As the pistons progress deeper into the cylinder bore, the brake fluid that is put under pressure transmits this force through both systems to friction pads at the wheels. A retaining ring fits into a groove near the end of the bore and holds the piston inside the cylinder.

Extra brake fluid is stored in two separate reservoirs. Most master cylinders are equipped with a fluid level warning switch. The switch causes the red brake warning lamp on the dash to turn on.

Vehicles with rear drum brakes have a residual check valve that keeps light (or residual) pressure in the brake lines and also at the wheel cylinders. Without residual pressure, air can be sucked past the wheel cylinder cups and into the wheel cylinders when the brake pedal is released quickly.

The design of the master cylinder used on a particular model car depends on the capacity of the reservoir and the displacement of the pistons at the wheel. Disc brakes require larger reservoirs than drum brakes because the caliper cylinder is much larger than that of a wheel cylinder. Some internal parts of the master cylinder may also change with application.

Hydraulic Tubes and Hoses

Steel tubing and flexible synthetic rubber hosing transmit brake fluid pressure from the master cylinder to the wheel cylinders and calipers of the drum and disc brakes. Fluid transfer from the master cylinder is usually routed through one or more valves and then into the steel tubing and hoses. Assorted fittings are used to connect steel tubing to junction blocks or other tubing sections. The most common fitting is the double or inverted flare style.

Brake hoses offer flexible connections to wheel units so that steering and suspension members can operate without damaging the brake system. Typical brake hoses are constructed of multiple layers of fabric impregnated with a synthetic rubber.

Hydraulic System Safety Switches and Valves

A pressure differential valve is used in all dual brake systems to operate a warning light switch. Its main purpose is to tell the driver if pressure is lost in either of the two hydraulic systems. Since each brake hydraulic system functions independently, it is possible that the driver might not notice immediately that pressure and braking are lost. When a pressure loss occurs, brake pedal travel increases and a more-than-usual effort is needed for braking. Should the driver not notice the extra effort needed, the warning light is actuated.

Metering and proportioning valves are used to balance the braking characteristics of disc and drum brakes. The braking response of the disc brakes is immediate and is directly proportionate to the effort applied at the pedal. Drum brake response is delayed because it takes some time for the hydraulic pressure to move the wheel cylinder pistons and overcome the force of their return springs and force the brake shoes to contact the drum.

A metering valve in the front brake line holds off pressure going to the front disc calipers. This delay allows pressure to build up in the rear drums first. When the rear brakes begin to take hold, the hydraulic pressure builds to the level needed to open the metering valve. When the metering valve opens, line pressure is high enough to operate the front discs. This process provides for better balance of the front and rear brakes. It also prevents lockup of the front brakes by keeping pressure from them until the rear brakes have started to operate.

The proportioning valve is used to control rear brake pressures, particularly during hard stops. When the pressure to the rear brakes reaches a specified level, the proportioning valve overcomes the force of its spring-loaded piston, stopping the flow of fluid to the rear brakes. By doing so, it regulates rear brake system pressure and adjusts for the difference in pressure between front and rear brake systems and keeps front and rear braking forces in balance.

A height-sensing proportional valve provides two different brake balance modes to the rear brakes based on vehicle load. This is accomplished by turning the valve on or off. When the vehicle is not loaded, hydraulic pressure is reduced to the rear brakes. When the vehicle is carrying a full load, the actuator lever moves up to change the valve's setting. The valve now allows full hydraulic pressure to the rear brakes. The valve is mounted to the frame above the rear axle. The valve has an actuator lever connected by a link to the lower shock absorber bracket. The valve is turned on and off as the axle to frame height changes due to load in the vehicle.

Most new cars have a combination valve in their hydraulic system. This valve is simply a single unit that combines the metering and proportioning valves with the pressure differential valve. There are two variations of the two-function combination valve. One variation does the proportioning valve and brake warning light switch functions. The other performs the metering valve and brake warning light switch functions.

The stop light (stop light/speed control) switch and mounting bracket assembly is attached to the brake pedal bracket and is activated by pressing the brake pedal. A mechanical stop light switch is operated by contact with the brake pedal or with a bracket attached to the pedal. A hydraulic switch is operated by hydraulic pressure developed in the master cylinder. In both types, the circuit through the switch is open when the brake pedal is released. When the brakes are applied, the circuit through the switch closes and causes the stoplights to come on.

Drum Brakes

A drum brake assembly consists of a cast-iron drum, which is bolted to and rotates with the vehicle's wheel, and a fixed backing plate to which are attached the shoes and other components, such as the wheel cylinders, automatic adjusters, and brake shoe linkages. Additionally, there might be some extra hardware for parking brakes. The shoes are surfaced with frictional linings, which contact the inside of the drum when the brakes are applied. The shoes are forced outward by hydraulic pistons located inside the wheel cylinder. As the drum rubs against the shoes, the energy of the moving drum is transformed into heat. This heat energy is passed into the atmosphere. When the brake pedal is released, hydraulic pressure drops, and the pistons are pulled back to their unapplied position by return springs.

Drum Brake Operation

Drum brake operation is fairly simple. The most important fact contributing to the effectiveness of a drum brake is the brake shoe pressure or force directed against the drum. With the vehicle moving with the brakes on, the applied force of the brake shoe, pressing against the brake drum, increasingly multiplies itself. This is because the brake's anchor pin acts as a brake shoe stop and prohibits the brake shoe from its tendency to follow the movement of the rotating drum. The result is a wedging action between the brake shoe and brake drum. The wedging action combined with the applied brake force creates a self-multiplied brake force.

Drum Brake Components

The backing plate provides a foundation for the brake shoes and associated hardware. The plate is secured and bolted to the axle flange or spindle. The wheel cylinder, under hydraulic pressure, forces the brake's shoes against the drum. There are also two linked brake shoes attached to the backing plate. The shoe must support the lining and carry it into the drum so that the pressure is distributed across the lining surface during brake application. Shoe return springs and shoe hold-down parts maintain the correct shoe position and clearance. Most drum brakes are self-adjusting. Others require manual adjustment mechanisms. Brake drums provide the rubbing surface area for the linings. Drums must withstand high pressures without excessive flexing and also dissipate large quantities of heat generated during brake application. Finally, the rear drum brakes on most vehicles include the parking brakes.

Wheel cylinders convert the hydraulic pressure supplied by the master cylinder into a mechanical force at the wheel brakes. The space in the wheel cylinder bore between the cups is filled with fluid. When the brake pedal is depressed, additional brake fluid is forced into the cylinder bore. The additional fluid, which is under pressure, moves the cups and pistons outward in the bore. This piston movement forces the shoe links and brake shoes outward to contact the drum and thus apply the brakes.

Each drum in the drum braking system contains a set of shoes. The primary (or leading) shoe is the one that is toward the front of the vehicle. The friction between the primary shoe and the brake drum forces the shoe to shift slightly in the direction that the drum is turning (an anchor pin permits just limited movement). The shifting of the primary shoe forces it against the bottom of the secondary shoe, which causes the secondary shoe to contact the drum. The secondary (or trailing) shoe is the one that is toward the rear of the vehicle. It comes into contact as a result of the movement and pressure from the primary shoe and wheel cylinder piston and increases the braking action.

The brake shoes are either held against the anchor by the shoe return springs or against the support plate pads by shoe hold-down springs. The shoes are linked together at the end opposite the anchor by an adjuster and a spring. The adjuster holds them apart. The spring holds them against the adjuster ends. Return springs can be separately hooked into a link or a guide or strung between the shoes.

Modern automotive brake drums are made of heavy cast iron (some are aluminum with an iron or steel sleeve or liner) with a machined surface inside against which the linings on the brake shoes generate friction when the brakes are applied. This results in the creation of a great deal of heat.

Drum Brake Designs

There are two brake designs in common use: duo-servo (or self-energizing) drum brakes and non-servo (or leading-trailing) drum brakes. The name duo-servo brake drum is derived from the fact that the self-energizing force is transferred from one shoe to the other with the wheel rotating in either direction.

When the brake shoes contact the rotating drum in either direction of rotation, they tend to move with the drum until one shoe contacts the anchor and the other shoe is stopped by the star wheel adjuster link. With forward rotation, frictional forces between the lining and the drum of the primary shoe result in a force acting on the adjuster link to apply the secondary shoe. This adjuster link force into the secondary shoe is many times greater than the wheel cylinder input force acting on the primary shoe. The force of the adjuster link into the secondary shoe is again multiplied by the frictional forces between the secondary lining and rotating drum, and all of the resultant force is taken on the anchor pin. In normal forward braking, the friction developed by the secondary lining is greater than the primary lining. Therefore, the secondary brake lining is usually thicker and has more surface area. The roles of the primary and secondary linings are reversed in braking the vehicle when backing up.

The non-servo drum brake system has no servo action. On forward brake applications, the forward (leading) shoe friction forces are developed by the pressure in the wheel cylinder. The shoe's friction forces work against the anchor pin at the bottom of the shoe. The trailing shoe is also actuated by wheel cylinder pressure, but can only support a friction force equal to the wheel cylinder piston force. The trailing shoe anchor pin supports no friction load. The leading shoe in this brake is energized and does most of the braking in comparison to the non-energized trailing shoe.

Automatic Brake Adjusters

Self-adjuster assemblies (whether cable, crank, or lever) are installed on one shoe and operated whenever the shoe moves away from its anchor. The upper link, or cable eye, is attached to the anchor. As the shoe moves, the cable pulls over a guide mounted on the shoe web (the crank or lever pivots on the shoe web) and operates a lever (pawl), which is attached to the shoe so that it engages a star wheel tooth. The pawl is located on the outer side of the star wheel and, on different styles, slightly above or below the wheel centerline so that it serves as a ratchet lock, which prevents the adjustment from backing off. However, whenever lining wears enough to permit sufficient shoe movement, brake application pulls the pawl high enough to engage the next tooth. As the brake is released, the adjuster spring returns the pawl, thus advancing the star wheel one notch.

The actual parts used to control self-adjusting vary with the type of drum brake system, as well as the manufacturer.

Drum Parking Brakes

The parking brake keeps a vehicle from rolling while it is parked. It is important to remember that the parking brake is not part of the vehicle's hydraulic braking system. It works mechanically, using a lever assembly connected through a cable system to the rear drum service brakes.

Parking brakes can be either hand or foot operated. The pedal or lever assembly is designed to latch into an applied position and is released by pulling a brake release handle or pushing a release button. On some vehicles, a vacuum power unit is connected by a rod to the upper end of the release lever. The vacuum motor is actuated to release the parking brake whenever the engine is running and the transmission is in forward driving gear.

The starting point of a typical parking brake cable and lever system is the foot pedal or hand lever. This assembly is a variable ratio lever mechanism that converts input effort of the operator and pedal/lever travel into output force with less travel. As the pedal is being depressed or lever raised, the front cable assembly tightens and tensile force is transmitted through the car's brake cable system to the rear brakes. This tension pulls the flexible steel cables attached to each of the rear brakes. It serves to operate the internal lever and strut mechanism of each rear brake, expanding the brake shoes against the drum. Springs return the shoes to the unapplied position when the parking brake is released and tensile forces in the cable system are relaxed.

An electronic switch, triggered when the parking brake is applied, lights the brake indicator in the instrument panel when the ignition is turned on. The light goes out when either the pedal or control is released or the ignition is turned off.

The cable routing system in a typical parking brake arrangement uses a three-lever setup to multiply the physical effort of the operator. First is the pedal assembly or handgrip. When moved it multiplies the operator's effect and pulls the front cable. The front cable, in turn, pulls the equalizer lever. The equalizer lever multiplies the effort of the pedal assembly, or handgrip, and pulls the rear cables equally. The equalizer allows the rear brake cables to slip slightly so as to balance out small differences in cable length or adjustment. The rear cables pull the parking brake levers in the drum brake assembly.

Disc Brakes

Disc brakes resemble the brakes on a bicycle: the friction elements are in the form of pads, which are squeezed or clamped about the edge of a rotating wheel. In automotive disc brakes, this wheel is a separate unit, called a rotor, and is mounted inboard of the vehicle's wheel. Since the pads clamp against both sides of it, both sides are machined smooth. Often the two surfaces are separated by a finned center section for better cooling (such rotors are called ventilated rotors). The pads are attached to metal shoes, which are actuated by pistons, the same as with drum brakes. The pistons are contained within a caliper assembly, a housing that wraps around the edge of the rotor.

The caliper is a housing containing the pistons and related seals, springs, and boots as well as the cylinders and fluid passages necessary to force the friction linings or pads against the rotor. The caliper resembles a hand in the way it wraps around the edge of the rotor. It is attached to the steering knuckle.

Calipers

A brake caliper converts hydraulic pressure into mechanical force. The housing contains the cylinder bore(s). In the cylinder bore is a groove that seats a square-cut seal. This groove is tapered toward the bottom of the bore to increase the compression on the edge of the seal that is nearest hydraulic pressure. The top of the cylinder bore is also grooved as a seat for the dust boot. A fluid inlet hole is machined into the bottom of the cylinder bore and a bleeder valve is located near the top of the casting.

A caliper can contain one, two, or four cylinder bores and pistons that provide uniform pressure distribution against the brake's friction pads. The pistons are relatively large in diameter and short in stroke to provide high pressure on the friction pad assemblies with a minimum of fluid displacement. The top of the pistons is grooved to accept the dust boot. The dust boot seats in a groove at the top of the cylinder bore and in a groove in the piston. The dust boot prevents moisture and road contamination from entering the bore.

The square-cut seal prevents fluid leakage between the cylinder bore wall and the piston. This rubber sealing ring also acts as a retracting mechanism for the piston when hydraulic pressure is released, causing the piston to return in its bore.

In addition, as the disc brake pads wear, the seal allows the piston to move further out to adjust automatically for the wear, without allowing fluid to leak and keeps the piston out and ready to clamp the surface of the rotor.

Disc brakes typically use fixed or floating calipers. There is also a sliding caliper, but its design is very similar to the floating caliper. The only difference is that sliding calipers slide on surfaces that have been machined smooth for this purpose, and floating calipers slide on special pins or bolts.

Fixed caliper disc brakes have a caliper assembly that is bolted in a fixed position and does not move when the brakes are applied. The pistons in both sides of the caliper come inward to force the pads against the rotor. A floating caliper has one hydraulic cylinder and a single piston. The caliper is attached to the spindle anchor plate with two threaded locating pins. A Teflon sleeve separates the caliper housing from each pin. The caliper slides back and forth on the pins as the brakes are actuated. When the brakes are applied, hydraulic pressure builds in the cylinder behind the piston and seal. Because hydraulic pressure exerts equal force in all directions, the piston moves evenly out of its bore.

The piston presses the inboard pad against the rotor. As the pad contacts the revolving rotor, greater resistance to outward movement is increased, forcing pressure to push the caliper away from the piston. This action forces the outboard pad against the rotor and both pads are applied with equal pressure.

With a sliding caliper assembly, the caliper slides or moves sideways when the brakes are applied. Unlike the floating caliper, the sliding caliper does not float on pins or bolts attached to the anchor plate. The sliding caliper has angular machined surfaces at each end that slide in mating machined surfaces on the anchor plate; this is where the caliper slides back and forth.

Some sliding calipers use a support key to locate and support the caliper in the anchor plate. The caliper support key is inserted between the caliper and the anchor plate.

Brake Pad Assembly

Brake pads are metal plates with the linings either riveted or bonded to them. Pads are placed at each side of the caliper and straddle the rotor. The inner brake pad, which is positioned against the piston, is not interchangeable with the outer brake pad.

Some brake pads have wear-sensing indicators. Audible sensors are thin, spring steel tabs that are riveted to or installed onto the edge of the pad's backing plate and are bent to contact the rotor when the lining wears down to a point that replacement is necessary. At that point, the sensor causes a high-pitched squeal at all times when the wheel is turning, except when the brakes are applied; then the noise goes away. The noise gives a warning to the driver that brake service is needed and perhaps saves the rotor from destruction.

Visual sensors inform the driver of the need for new linings. This method employs electrical contacts recessed in the pads that touch the rotor when the linings are worn out. This completes a circuit and turns on a dashboard warning light. Tactile sensors create pedal pulsation as the sensor on the rotor face contracts the sensor attached to the lower portion of the disc pad.

Hub and Rotor Assembly

The typical rotor is solid or ventilated and made of cast iron. Iron has a high coefficient of friction and withstands wear exceptionally well. Composite style rotors are growing in popularity and are found on many of today's FWD and four-wheel disc vehicles. The rotor is attached to and rotates with the wheel hub assembly.

A splash shield protects the rotor and pads from road splashes and dirt. It is also shaped to channel the flow of air over the exposed rotor surfaces. As long as the car is moving, this flow of air helps to cool the rotor.

Rear Disc/Drum (Auxiliary Drum) Parking Brake

The rear disc/drum or auxiliary drum parking brakc arrangement is found on some vehicles. On these brakes, the inside of each rear wheel hub and rotor assembly is used as the parking brake drum. The drum brake is a smaller version of a drum brake and is serviced like any other drum brake.

Rear Disc Parking Brakes

Instead of using an auxiliary drum and shoes to hold the vehicle when parked, these brakes have a mechanism that forces the pads against the rotor mechanically. One method for doing this is the ball-and-ramp arrangement. Another method that is found on many vehicles has a threaded, spring-loaded pushrod. As the parking brakes are applied, a mechanism rotates or unscrews the pushrod, which in turn pushes the piston out. Other types of actuating systems for rear disc brakes include the use of a variety of cams.

Vacuum-Assist Power Brakes

All vacuum-assisted power brake units are similar in design. They generate application energy by opposing engine vacuum to atmospheric pressure. A piston and cylinder, flexible diaphragm or bellows use this energy to provide braking assistance.

Vacuum-assist units are vacuum-suspended systems, which means the diaphragm inside the unit is balanced using engine vacuum until the brake pedal is depressed. Applying the brake allows atmospheric pressure to unbalance the diaphragm and allows it to move, generating application pressure. Vacuum boosters may be single diaphragm or tandem diaphragm.

Single diaphragm boosters have a vacuum power section that includes a front and rear shell, a power diaphragm, a return spring, and a pushrod. A control valve is an integral part of the power diaphragm and is connected through a valve rod to the brake pedal. It controls the degree of brake application or release in accordance with the pressure applied to the brake pedal.

When the brakes are applied, the valve rod and plunger move the power diaphragm. This action closes the control valve's vacuum port and opens the atmospheric port to admit air through the valve at the rear diaphragm chamber. With vacuum in the rear chamber, a force develops that moves the power diaphragm, hydraulic pushrod, and hydraulic piston or pistons to close the compensating port or ports and force fluid under pressure through the residual check valve or valves and lines into the front and rear brake assemblies.

As pressure develops in the master cylinder, a counter force acts through the hydraulic pushrod and reaction disc against the vacuum power diaphragm and valve plunger. This force tends to close the atmospheric port and reopen the vacuum port. Since this force is in opposition to the force applied to the brake pedal by the operator, it gives the operator a feel for the amount of brake applied.

Hydraulic-Assist Power Brakes

Hydraulic-assist power brakes use fluid pressure from the power steering pump, not vacuum, to help apply the brakes. The power steering pump provides a continuous flow of fluid to the brake booster whenever the engine is running. Three flexible hoses route the power steering fluid to the booster. One hose supplies pressurized fluid from the pump. Another hose routes the pressurized fluid from the booster to the power steering gear assembly. The third hose returns fluid from the booster to the power steering pump.

Some systems have a nitrogen charged pneumatic accumulator on the booster to provide reserve power assist pressure. If power steering pump pressure is not available, due to belt failure or similar problems, the accumulator pressure is used to provide brake assist.

The booster assembly consists of an open center spool valve and sleeve assembly, a lever assembly, an input rod assembly, a power piston, an output pushrod, and the accumulator. The booster assembly is mounted on the vehicle in much the same manner as a vacuum booster. The pedal rod is connected at the booster input rod end.

When the brake pedal is depressed, the pedal's pushrod moves the master cylinder's primary piston forward. This causes the lever assembly of the booster to move a sleeve forward to close off the holes leading to the open center of the spool valve. A small additional lever movement moves the spool valve into the spool valve bore. The spool valve then diverts some hydraulic fluid into the cavity behind the booster piston, building up hydraulic pressure. This hydraulic pressure moves the piston and a pushrod forward. The output pushrod moves the primary and secondary master cylinder pistons that apply pressure to the brake system. When the brake pedal is released, the spool and sleeve assemblies return to their normal position. Excess fluid behind the piston returns to the power steering pump reservoir through the return line. After the brakes have been released, pressurized fluid from the power steering pump flows into the booster through the open center of the spool valve and back to the power steering pump.

Antilock Brake Systems

Antilock brake systems (ABS) can be thought of as electronic/hydraulic pumping of the brakes for straight-line stopping under panic conditions. Good drivers have always pumped the brake pedal during panic stops to avoid wheel lockup and the loss of steering control. Antilock brake systems simply get the pumping job done much faster and in a much more precise manner than the fastest human foot. Keep in mind that a tire on the verge of slipping produces more friction with respect to the road than one that is locked and skidding. Once a tire loses its grip, friction is reduced and the vehicle takes longer to stop.

When the driver quickly and firmly applies the brakes and holds the pedal down, the brakes of a vehicle not equipped with ABS will almost immediately lock the wheels. The vehicle slides rather than rolls to a stop. During this time, the driver also has a very difficult time keeping the vehicle straight and the vehicle will skid, out of control. The locking of the wheels caused the skidding and lack of control. If the driver was able to release the brake pedal just before the wheels locked up and then reapply the brakes, the skidding could be avoided.

This release and apply of the brake pedal is exactly what an antilock system does. When the brake pedal is pumped or pulsed, pressure is quickly applied and released at the wheels. This is called pressure modulation. Pressure modulation works to prevent wheel locking. Antilock brake systems can modulate the pressure to the brakes as often as 15 times per second. By modulating the pressure to the brakes, friction between the tires and the road is maintained and the vehicle is able to come to a controllable stop.

ABS precisely controls the slip rate of the wheels to ensure maximum grip force from the tires, and thereby ensures the maneuverability and stability of the vehicle. An ABS control module calculates the slip rate of the wheels based on the vehicle speed and the speed of the wheels, and then it controls the brake fluid pressure to attain the target slip rate.

ABS Hydraulic Components

An accumulator is used to store hydraulic fluid to maintain high pressure in the brake system and to provide residual pressure for power-assisted braking. Normally the accumulator is charged with nitrogen gas and is an integral part of the modulator unit. This unit is typically found on vehicles with a hydraulically assisted brake system.

An antilock hydraulic control valve assembly controls the release and application of brake system pressure to the wheel brake assemblies. It may be an integral type, meaning this unit is combined with the power-boost and master cylinder units into one assembly or it may a non-integral type. Non-integral units are mounted externally from the master cylinder/power booster unit and are located between the master cylinder and wheel brake assemblies. Both types generally contain solenoid valves that control the releasing, the holding, and the applying of brake system pressure.

The booster pump is an assembly of an electric motor and pump. The booster pump provides pressurized fluid for the ABS system. The system's control unit controls the pump's motor. The booster pump is also called the electric pump and motor assembly.

The booster/master cylinder assembly, sometimes referred to as the hydraulic unit, contains the valves and pistons needed to modulate hydraulic pressure in the wheel circuits during ABS operation. Power brake assist is provided by pressurized brake fluid supplied by a hydraulic pump.

Fluid accumulators temporarily store the brake fluid that is removed from the wheel brake units during an ABS cycle. The fluid is then used by the pump to build pressure for the brake hydraulic system. There are normally two fluid accumulators in a hydraulic control unit, one for each of the primary and secondary hydraulic circuits.

The hydraulic control unit contains the solenoid valves, fluid accumulators, pump, and an electric motor. This is actually a combination unit of many individual components found separately in some systems. The unit may have one pump and one motor or it may have one motor and two pumps: one pump for half of the hydraulic system and a pump for the other half.

The main valve is a two-position valve controlled by the ABS control module and is open only in the ABS mode. When open, pressurized brake fluid from the booster circuit is directed into the master cylinder circuits to prevent excessive pedal travel.

A modulator unit controls the flow of pressurized brake fluid to the individual wheel circuits. Normally the modulator is made up of solenoids that open and close valves, several valves that control the flow of fluid to the wheel brake units, and electrical relays that activate or deactivate the solenoids through the commands of the control module. The control module switches the solenoids on or off to increase, decrease, or maintain the hydraulic pressure to the individual wheel units.

Two solenoid valves are used to control each circuit or channel. One controls the inlet valve of the circuit, the other the outlet valve. When inlet and outlet valves of a circuit are used in combination, pressure can be increased, decreased, or held steady in the circuit. The control module determines the position of each valve. Outlet valves are normally closed, and inlet valves are normally open.

Electrical/Electronic ABS Components

The ABS control module is normally mounted inside the trunk on the wheel housing, mounted to the master cylinder, or is part of the hydraulic control unit, PCM, or BCM. It monitors system operation and controls antilock function when needed. The module relies on inputs from the wheel speed sensors and feedback from the hydraulic unit to determine if the antilock brake system is operating correctly and to determine when the antilock mode is required.

The antilock brake pedal sensor switch is normally closed. When the brake pedal travel exceeds the switch's setting during an antilock stop, the switch opens and the control module grounds the pump motor relay coil. This energizes the relay and turns the pump motor on. When the pump motor is running, the hydraulic reservoir is filled with high-pressure brake fluid, and the brake pedal will be pushed up until the antilock brake pedal sensor switch closes. When the switch closes, the pump motor is turned off and the brake pedal will drop some during each ABS control cycle until the switch opens again and the pump motor is turned on again. This minimizes pedal feedback during ABS cycling.

Most ABS-equipped vehicles are fitted with two different warning lights. One of the warning lights is tied directly to the ABS system, whereas the other lamp is part of the base brake system. All vehicles have a RED warning light. This lamp is lit when there is a problem with the brake system or when the parking brake is on. An AMBER warning lamp lights when there is a fault in the ABS system.

A lateral acceleration sensor is used to monitor the sideward movement of the vehicle while it is turning a corner. This information is sent to the control module to ensure proper braking during turns.

Pressure switches are used to control the operation of the pump motor and the low-pressure warning light circuit. The pressure switch grounds the pump motor relay coil circuit, activating the pump when accumulator pressure drops below a specified level. The switch also turns off the motor when the pressure reaches a high limit.

The pressure differential switch is located in the modulator unit and sends a signal to the control module whenever there is an undesirable difference in hydraulic pressures within the brake system.

A toothed ring can be located on an axle shaft, differential gear, or a wheel's hub. This ring is used in conjunction with the wheel speed sensor. The ring has a number of teeth around its circumference. As the ring rotates and each tooth passes by the wheel-speed sensor, an AC voltage signal is generated between the sensor and the tooth. As the tooth moves away from the sensor, the signal is broken until the next tooth comes close to the sensor. The end result is a pulsing signal that is sent to the control module. The control module translates the signal into wheel speed. The toothed ring may also be called the reluctor, tone ring, or gear pulser.

Types of Antilock Brake Systems

In addition to being classified as integral and non-integral antilock brake systems, systems can be broken down into the level of control they provide. Antilock brake systems can be 1-, 2-, 3-, or 4-channel two- or four-wheel systems. A channel is merely a hydraulic circuit to the brakes.

Two-wheel systems offer antilock brake performance to the rear wheels only. They do not provide antilock performance to the steering wheels. These systems can be either 1- or 2-channel systems. In 1-channel systems, the rear brakes on both sides of the vehicle are modulated at the same time to control skidding. These systems rely on the input from a centrally located speed sensor. The speed sensor is normally positioned on the ring gear in the differential unit, transmission, or transfer case. A 2-channel setup can be used to modulate the pressure to each of the rear wheels independently of each other. Modulation is controlled by the speed variances recorded by speed sensors located at each wheel.

Some hydraulic systems that are split from front to rear use a 3-channel circuit and are called four-wheel antilock brake systems. These systems have individual hydraulic circuits to each of the two front wheels, and a single circuit to the two rear wheels.

The antilock system that is the most effective and most common is a 4-channel system, in which sensors monitor each of the four wheels. With this continuous information, the ABS control module ensures that each wheel receives the exact braking force it needs to maintain both antilock and steering control.

Automatic Traction Control

Traction control systems rely on the technology and hardware of antilock braking systems to control tire traction and vehicle stability. Automatic traction control systems apply the brakes when a drive wheel attempts to spin and lose traction.

During operation, the system uses an electronic control module to monitor the wheel-speed sensors. If a wheel enters a loss-of-traction situation, the module applies braking force to the wheel in trouble. If there is a loss of traction, the speed of the wheel will be greater than expected for the particular vehicle speed. Wheel spin is normally limited to a 10 percent slippage.

Automatic Stability Control

Various stability control systems are found on today's vehicles. Like traction control systems, stability controls are based on and linked to the antilock brake system. On some vehicles the stability control system is also linked to the electronic suspension system.

Stability control systems momentarily apply the brakes at any one wheel to correct oversteer or understeer. The control unit receives signals from the typical sensors plus a yaw, lateral acceleration (G-force), and a steering angle sensor.

The system uses the angle of the steering wheel and the speed of the four wheels to calculate the path chosen by the driver. It then looks at lateral G-forces and vehicle yaw to measure where the vehicle is actually going.

SAFETY

In an automotive repair shop, there is great potential for serious accidents, simply because of the nature of the business and the equipment used. When people are careless, the automotive repair industry can be one of the most dangerous occupations. But, the chances of your being injured while working on a car are close to nil if you learn to work safely and use common sense. Safety is the responsibility of everyone in the shop.

Personal Protection

Some procedures, such as grinding, result in tiny particles of metal and dust that are thrown off at very high speeds. These metal and dirt particles can easily get into your eyes, causing scratches or cuts on your eyeball. Pressurized gases and liquids escaping a ruptured hose or hose-fitting can spray a great distance. If these chemicals get into your eyes, they can cause blindness. Dirt and sharp bits of corroded metal can easily fall down into your eyes while you are working under a vehicle.

Eye protection should be worn whenever you are exposed to these risks. To be safe, you should wear safety glasses whenever you are working in the shop. Some procedures may require that you wear other eye protection in addition to safety glasses. For example, when cleaning parts with a pressurized spray, you should wear a face shield. The face shield not only gives added protection to your eyes but also protects the rest of your face.

If chemicals such as battery acid, fuel, or solvents get into your eyes, flush them continuously with clean water. Have someone call a doctor and get medical help immediately.

Your clothing should be well fitted and comfortable but made of strong material. Loose, baggy clothing can easily get caught in moving parts and machinery. Some technicians prefer to wear coveralls or shop coats to protect their personal clothing. Your work clothing should offer you some protection but should not restrict your movement.

Long hair and loose, hanging jewelry can create the same type of hazard as loose-fitting clothing. They can get caught in moving engine parts and machinery. If you have long hair, tie it back or tuck it under a cap.

Never wear rings, watches, bracelets, and neck chains. These can easily get caught in moving parts and cause serious injury.

Always wear shoes or boots of leather or similar material with non-slip soles. Steel-tipped safety shoes can give added protection to your feet. Jogging or basketball shoes, street shoes, and sandals are inappropriate in the shop.

Good hand protection is often overlooked. A scrape, cut, or burn can limit your effectiveness at work for many days. A well-fitted pair of heavy work gloves should be worn during operations such as grinding and welding or when handling hot components. Always wear approved rubber gloves when handling strong and dangerous caustic chemicals.

Many technicians wear thin, surgical-type latex gloves whenever they are working on vehicles. These offer little protection against cuts but do offer protection against disease and grease buildup under and around your fingernails. These gloves are comfortable and are quite inexpensive.

A brake technician should always wear a respirator. This can be a simple throwaway mask, or a mask with replaceable filter cartridges. Put the respirator on when you remove the wheels from the car and keep it on until you are finished working.

The face piece of the respirator must seal against your face. Felt filters use an electrostatic charge to help trap dust. High-efficiency paper filters use a fiberglass paper with fine pores to stop the passage of dust.

The risk of asbestos exposure is highest when grinding the lining for brake shoes. Technicians seldom do this today, but if you must grind the lining, make sure you wear a respirator. The grinding area should have its own exhaust ventilation system to draw away the dust and fibers.

Accidents can be prevented simply by the way you act. The following are some guidelines to follow while working in a shop. This list does not include everything you should or shouldn't do; it merely presents some things to think about.

- Never smoke while working on a vehicle or while working with any machine in the shop.
- Playing around is not fun when it sends someone to the hospital.
- To prevent serious burns, keep your skin away from hot metal parts such as the radiator, exhaust manifold, tailpipe, catalytic converter, and muffler.

- Always disconnect electric engine cooling fans when working around the radiator. Many of these will turn on without warning and can easily chop off a finger or hand. Make sure you reconnect the fan after you have completed your repairs.
- When working with a hydraulic press, make sure the pressure is applied in a safe manner. It is generally wise to stand to the side when operating the press.
- Properly store all parts and tools by putting them away in a place where people will not trip over them. This practice not only cuts down on injuries, it also reduces time wasted looking for a misplaced part or tool.

Work Area Safety

Your entire work area should be kept clean and safe. Any oil, coolant, or grease on the floor can make it slippery. To clean up oil, use commercial oil absorbent. Keep all water off the floor. Water makes smooth floors slippery and is dangerous as a conductor of electricity. Aisles and walkways should be kept clean and wide enough to allow easy movement. Make sure the work areas around machines are large enough to allow the machinery to be operated safely.

Gasoline is a highly flammable volatile liquid. Something that is flammable catches fire and burns easily. A volatile liquid is one that vaporizes very quickly. Flammable volatile liquids are potential firebombs. Always keep gasoline or diesel fuel in an approved safety can and never use gasoline to clean your hands or tools.

Handle all solvents (or any liquids) with care to avoid spillage. Keep all solvent containers closed, except when pouring. Proper ventilation is very important in areas where volatile solvents and chemicals are used. Solvent and other combustible materials must be stored in approved and designated storage cabinets or rooms with adequate ventilation. Never light matches or smoke near flammable solvents and chemicals, including battery acids.

Oily rags should also be stored in an approved metal container. When these oily, greasy, or paint-soaked rags are left lying about or are not stored properly, they can cause spontaneous combustion. Spontaneous combustion results in a fire that starts by itself, without a match.

Disconnecting the vehicle's battery before working on the electrical system, or before welding, can prevent fires caused by a vehicle's electrical system. To disconnect the battery, remove the negative or ground cable from the battery and position it away from the battery.

Know where all of the shop's fire extinguishers are located. Fire extinguishers are clearly labeled as to what type they are and what types of fire they should be used on. Make sure you use the correct type of extinguisher for the type of fire you are dealing with. A multipurpose dry chemical fire extinguisher will put out ordinary combustibles, flammable liquids, and electrical fires. Never put water on a gasoline fire because it will just cause the fire to spread. The proper fire extinguisher will smother the flames.

During a fire, never open doors or windows unless it is absolutely necessary; the extra draft will only make the fire worse. Make sure the fire department is contacted before or during your attempt to extinguish a fire.

Air Bag Safety

When service is performed on any air bag system component, always disconnect the negative battery cable, isolate the cable end, and wait for the amount of time specified by the vehicle manufacturer before proceeding with the necessary diagnosis or service. The average waiting period is two minutes, but some vehicle manufacturers specify up to ten minutes. Failure to observe this precaution may cause accidental air bag deployment and personal injury.

Replacement air bag system parts must have the same part number as the original part. Replacement parts of lesser or questionable quality must not be used. Improper or inferior components may result in inappropriate air bag deployment and injury to the vehicle occupants.

Do not strike or jar a sensor or an air bag system diagnostic monitor (ASDM). This may cause air bag deployment or make the sensor inoperative. Accidental air bag deployment may cause personal injury, and an inoperative sensor may result in air bag deployment failure, causing personal injury to vehicle occupants.

All sensors and mounting brackets must be properly torqued to ensure correct sensor operation before an air bag system is powered up. If sensor fasteners do not have the proper torque, improper air bag deployment may result in injury to vehicle occupants.

When working on the electrical system on an air-bag-equipped vehicle, use only the vehicle manufacturer's recommended tools and service procedures. The use of improper tools or service procedures may cause accidental air bag deployment and personal injury. For example, do not use 12V or self-powered test lights when servicing the electrical system on an air-bag-equipped vehicle.

Brake Cleaning Solvents

The various cleaning solvents used on brake systems must be handled carefully. The vapors from these solvents can cause drowsiness or a loss of consciousness. Very high levels of exposure can be fatal.

Brake Fluid Safety

Used brake fluid is considered a hazardous waste and must be disposed of according to local, state, and federal laws. DOT 3 and DOT 4 fluids will damage the finish of the vehicle. Always wipe up any spilled fluid immediately. They are also hygroscopic, which means they attract water and will absorb moisture out of the air. Water mixed with brake fluid lowers its boiling point and reduces its effectiveness. This can lead to unsafe brake system operation. Always keep brake fluid containers tightly closed when not in use.

Brake fluids will also absorb the oils from your skin and lead to dryness, irritation, and inflammation. Wear gloves when handling brake fluid and immediately wash off any fluid that contacts your skin.

ABS Hydraulic Pressure Safety

Many antilock brake systems generate extremely high brake fluid pressures. Failure to depressurize the hydraulic accumulator before servicing any part of the system could cause severe injury. Follow the exact shop manual procedure for the vehicle being serviced.

Hydraulic Power-Assisted Brakes

Pressures in brake systems with hydraulic power-assist units are also very high. These units rely on the power steering pump and an accumulator to store pressurized brake fluid. The way to relieve this pressure is similar to that for ABS. Follow the exact shop manual procedure for the vehicle being serviced.

Tool and Equipment Safety

Careless use of simple hand tools such as wrenches, screwdrivers, and hammers causes many shop accidents that could be prevented. Keep all hand tools grease-free and in good condition. Tools that slip can cause cuts and bruises. If a tool slips and falls into a moving part, it can fly out and cause serious injury.

Use the proper tool for the job. Make sure the tool is of professional quality. Using poorly made tools or the wrong tools can damage parts or the tool itself, or could cause injury. Never use broken or damaged tools.

Safety around power tools is very important. Serious injury can result from carelessness. Always wear safety glasses when using power tools. If the tool is electrically powered, make sure it is properly grounded. Before using it, check the wiring for cracks in the insulation, as well as for bare wires. Also, when using electrical power tools, never stand on a wet or damp floor. Never leave a running power tool unattended.

Tools that use compressed air are called pneumatic tools. Compressed air is used to inflate tires, apply paint, and drive tools. Compressed air can be dangerous when it is not used properly.

When using compressed air, safety glasses and/or a face shield should be worn. Particles of dirt and pieces of metal, blown by the high-pressure air, can penetrate your skin or get into your eyes.

Before using a compressed air tool, check all hose connections. Always hold an air nozzle or air control device securely when starting or shutting off the compressed air. A loose nozzle can whip suddenly and cause serious injury. Never point an air nozzle at anyone. Never use compressed air to blow dirt from your clothes or hair. Never use compressed air to clean the floor or workbench.

Always be careful when raising a vehicle on a lift or a hoist. Adapters and hoist plates must be positioned correctly to prevent damage to the underbody of the vehicle. There are specific lift points that allow the weight of the vehicle to be evenly supported by the adapters or hoist plates. The correct lift points can be found in the vehicle's service manual. Before operating any lift or hoist, carefully read the operating manual and follow the operating instructions.

Once you feel the lift supports are properly positioned under the vehicle, raise the lift until the supports contact the vehicle. Then, check the supports to make sure they are in full contact with the vehicle. Shake the vehicle to make sure it is securely balanced on the lift, and then raise the lift to the desired working height. Before working under a car, make sure the lift's locking devices are engaged.

A vehicle can be raised off the ground by a hydraulic jack. The jack's lifting pad must be positioned under an area of the vehicle's frame or at one of the manufacturer's recommended lift points. Never place the pad under the floor pan or under steering and suspension components because they are easily damaged by the weight of the vehicle. Always position the jack so the wheels of the vehicle can roll as the vehicle is being raised.

Safety stands, also called jack stands, should be placed under a sturdy chassis member, such as the frame or axle housing, to support the vehicle after it has been raised by a jack. Once the safety stands are in position, the hydraulic pressure in the jack should be slowly released until the weight of the vehicle is on the stands. Never move under a vehicle when it is supported only by a hydraulic jack. Rest the vehicle on the safety stands before moving under the vehicle.

Heavy parts of the automobile, such as engines, are removed with chain hoists or cranes. Cranes often are called cherry pickers. To prevent serious injury, chain hoists and cranes must be properly attached to the parts being lifted. Always use bolts with enough strength to support the object being lifted. After you have attached the lifting chain or cable to the part that is being removed, have your instructor check it. Place the chain hoist or crane directly over the assembly. Then, attach the chain or cable to the hoist.

Parts cleaning is a necessary step in most repair procedures. Always wear the appropriate protection when using chemical, abrasive, and thermal cleaners.

Vehicle Operation

When the customer brings a vehicle in for service, certain driving rules should be followed to ensure your safety and the safety of those working around you. For example, before moving a car into the shop, buckle your safety belt. Make sure no one is near, the way is clear, and there are no tools or parts under the car before you start the engine. Check the brakes before putting the vehicle in gear. Then, drive slowly and carefully in and around the shop.

If the engine must be running while you are working on the car, block the wheels to prevent the car from moving. Place the transmission into park for automatic transmissions or into neutral for manual transmissions. Set the parking (emergency) brake. Never stand directly in front of or behind a running vehicle.

Run the engine only in a well-ventilated area to avoid the danger of poisonous carbon monoxide (CO) in the engine exhaust. CO is an odorless but deadly gas. Most shops have an exhaust ventilation system and you should always use it. Connect the hose from the vehicle's tailpipe to the intake for the vent system. Make sure the vent system is turned on before running the engine. If the work area does not have an exhaust venting system, use a hose to direct the exhaust out of the building.

HAZARDOUS MATERIALS AND WASTES

A typical shop contains many potential health hazards for those working in it. These hazards can cause injury, sickness, health impairments, discomfort, and even death. Here is a short list of the different classes of hazards:

- Chemical hazards are caused by high concentrations of vapors, gases, or solids in the form of dust.
- Hazardous wastes are those substances that result from performing a service.
- Physical hazards include excessive noise, vibration, pressures, and temperatures.
- Ergonomic hazards are conditions that impede normal and/or proper body position and motion.

There are many government agencies charged with ensuring safe work environments for all workers. These include the Occupational Safety and Health Administration (OSHA), Mine Safety and Health Administration (MSHA), and National Institute for Occupational Safety and Health (NIOSH). These, as well as state and local governments, have instituted regulations that must be understood and followed. Everyone in a shop has the responsibility for adhering to these regulations.

An important part of a safe work environment is the employees' knowledge of potential hazards. Right-to-know laws concerning all chemicals protect every employee in the shop. The general intent of right-to-know laws is to ensure that employers provide their employees with a safe working place as far as hazardous materials are concerned.

All employees must be trained about their rights under the legislation, the nature of the hazardous chemicals in their workplace, and the contents of the labels on the chemicals. All of the information about each chemical must be posted on material safety data sheets (MSDS) and must be accessible. The manufacturer of the chemical must give these sheets to its customers, if they are requested to do so. They detail the chemical composition and precautionary information for all products that can present a health or safety hazard.

Employees must become familiar with the general uses, protective equipment, accident or spill procedures, and any other information regarding the safe handling of the hazardous material. This training must be given to employees annually and provided to new employees as part of their job orientation.

A hazardous material must be properly labeled, indicating what health, fire, or reactivity hazard it poses and what protective equipment is necessary when handling each chemical. The manufacturer of the hazardous materials must provide all warnings and precautionary information, which must be read and understood by the user before use. A list of all hazardous materials used in the shop must be posted for the employees to see.

Shops must maintain documentation on the hazardous chemicals in the workplace, proof of training programs, records of accidents or spill incidents, satisfaction of employee requests for specific chemical information via the MSDS, and a general right-to-know compliance procedure manual utilized within the shop.

When handling any hazardous materials or hazardous waste, make sure you follow the required procedures for handling such material. Also wear the proper safety equipment listed on the MSDS. This includes the use of approved respirator equipment.

Some of the common hazardous materials that automotive technicians use are: cleaning chemicals, fuels (gasoline and diesel), paints and thinners, battery electrolyte (acid), used engine oil, refrigerants, and engine coolant (anti-freeze).

Many repair and service procedures generate what are known as hazardous wastes. Dirty solvents and cleaners are good examples of hazardous wastes. Something is classified as a hazardous waste if it is on the EPA list of known harmful materials or has one or more of the following characteristics.

- *Ignitability.* If it is a liquid with a flash point below 140°F or a solid that can spontaneously ignite.
- *Corrosivity.* If it dissolves metals and other materials or burns the skin.
- *Reactivity.* Any material that reacts violently with water or other materials or releases cyanide gas, hydrogen sulfide gas, or similar gases when exposed to low pH acid solutions. This also includes material that generates toxic mists, fumes, vapors, and flammable gases.
- *Toxicity.* Materials that leach one or more of eight heavy metals in concentrations greater than 100 times primary drinking water standard concentrations.

Complete EPA lists of hazardous wastes can be found in the Code of Federal Regulations. It should be noted that no material is considered hazardous waste until the shop is finished using it and is ready to dispose of it.

The following list covers the recommended procedure for dealing with some of the common hazardous wastes. Always follow these and any other mandated procedures.

Oil Recycle oil. Set up equipment, such as a drip table or screen table with a used oil collection bucket, to collect oils dripping off parts. Place drip pans underneath vehicles that are leaking fluids onto the storage area. Do not mix other wastes with used oil, except as allowed by your recycler. Used oil generated by a shop (and/or oil received from household "do-it-yourself" generators) may be burned on site in a commercial space heater. Also, used oil may be burned for energy recovery. Contact state and local authorities to determine requirements and to obtain necessary permits.

Oil Filters Drain for at least 24 hours, crush, and recycle used oil filters.

Batteries Recycle batteries by sending them to a reclaimer or back to the distributor. Keeping shipping receipts can demonstrate that you have done the recycling. Store batteries in a watertight, acid-resistant container. Inspect batteries for cracks and leaks when they come in. Treat a dropped battery as if it were cracked. Acid residue is hazardous because it is corrosive and may contain lead and other toxic substances. Neutralize spilled acid, by using baking soda or lime, and dispose of it as a hazardous material.

Metal residue from machining Collect metal filings when machining metal parts. Keep them separate and recycle if possible. Prevent metal filings from falling into a storm sewer drain.

Refrigerants Recover and/or recycle refrigerants during the servicing and disposal of motor vehicle air conditioners and refrigeration equipment. It is not allowable to knowingly vent refrigerants to the atmosphere. Recovering and/or recycling during servicing must be performed by an EPA-certified technician using certified equipment and following specified procedures.

Solvents Replace hazardous chemicals with less toxic alternatives that perform equally. For example, substitute water-based cleaning solvents for petroleum-based solvent degreasers. To reduce the amount of solvent used when cleaning parts, use a two-stage process: dirty solvent followed by fresh solvent. Hire a hazardous waste management service to clean and recycle solvents. (Some spent solvents must be disposed of as hazardous waste, unless recycled properly). Store solvents in closed containers to prevent evaporation. Evaporation of solvents contributes to ozone depletion and smog formation. In addition, the residue from evaporation must be treated as a hazardous waste. Properly label spent solvents and store on drip pans or in diked areas and only with compatible materials.

Containers Cap, label, cover, and properly store aboveground outdoor liquid containers and small tanks within a diked area and on a paved impermeable surface to prevent spills from running into surface or ground water.

Other solids Store materials such as scrap metal, old machine parts, and worn tires under a roof or tarpaulin to protect them from the elements and to prevent the possibility of creating contaminated runoff. Consider recycling tires by retreading them.

Liquid recycling Collect and recycle coolants from radiators. Store transmission fluids, brake fluids, and solvents containing chlorinated hydrocarbons separately, and recycle or dispose of them properly.

Shop towels or rags Keep waste towels in a closed container marked "contaminated shop towels only." To reduce costs and liabilities associated with disposal of used towels (which can be classified as hazardous wastes), investigate using a laundry service that is able to treat the wastewater generated from cleaning the towels.

Waste storage Always keep hazardous waste separate, properly labeled, and sealed in the recommended containers. The storage area should be covered and may need to be fenced and locked if vandalism could be a problem. Select a licensed hazardous waste hauler after seeking recommendations and reviewing the firm's permits and authorizations.

Asbestos Health Hazards

One of the greatest safety concerns in any shop doing brake service is exposure of personnel to asbestos dust. Although asbestos is no longer used as a lining material, vehicles are still on the road with asbestos linings. And you definitely need to be concerned about exposure to asbestos. If inhaled, asbestos dust from brake and clutch linings is a serious health hazard. Once inhaled, the fine asbestos fibers stay in your lungs and can cause asbestosis and lung cancer.

The best way to avoid asbestos dust is not to create it. Handle brake parts carefully. Never use compressed air or dry brushing for cleaning dust from brake drums, brake backing plates, and brake assemblies. Use specialized vacuum cleaning equipment or the proper wet-cleaning procedures. Keep yourself and your clothing clean. Wash thoroughly before eating, and avoid eating in the shop area.

Vacuum cleaner bags from asbestos vacuum cleaners and any cleaning rags that get contaminated with asbestos should be disposed of according to EPA or OSHA standards. In general, these materials must be sealed in plastic bags with proper warnings on them. The bags must be disposed of following proper procedures for hazardous waste.

A clean brake system service area is essential to technician health. Any brake dust in the shop must be vacuumed up with the correct type of EPA-approved vacuum systems. These vacuums have multistage filters to prevent the asbestos from getting out of the vacuum cleaner. Never dry sweep or blow brake dust off brake parts, equipment, or the shop floor.

The HEPA (high efficiency particulate air) filter vacuum system can also be used for general floor cleanup. If a vacuum is not available, mop the shop floor with water to collect the asbestos dust.

You can also protect yourself and your family by leaving work clothes that could be contaminated with asbestos at work. If possible, shower at work or shower as soon as you get home to remove any asbestos from your skin.

Never drink or smoke in an area where brake work is being performed. Wash thoroughly before eating, drinking, or smoking.

NATEF TASK LIST FOR BRAKE SYSTEMS

A. General Brake Systems Diagnosis

	Task	Priority
A.1.	Identify and interpret brake system concern: determine necessary action.	Priority Rating 1
A.2.	Research applicable vehicle and service information, such as brake system operation, vehicle service history, service precautions, and technical service bulletins.	Priority Rating 1
A.3.	Locate and interpret vehicle and major component identification numbers (VIN, vehicle certification labels, calibration labels).	Priority Rating 1

B. Hydraulic System Diagnosis and Repair

	Task	Priority
B.1.	Diagnose pressure concerns in the brake system using hydraulic principles (Pascal's Law).	Priority Rating 1
B.2.	Measure brake pedal height; determine necessary action.	Priority Rating 2
B.3.	Check master cylinder for internal and external leaks and proper operation; determine necessary action.	Priority Rating 2
B.4.	Remove, bench bleed, and reinstall master cylinder.	Priority Rating 1
B.5.	Diagnose poor stopping, pulling or dragging concerns caused by malfunctions in the hydraulic system; determine necessary action.	Priority Rating 1
B.6.	Inspect brake lines, flexible hoses, and fittings for leaks, dents, kinks, rust, cracks, bulging or wear; tighten loose fittings and supports; determine necessary action.	Priority Rating 2
B.7.	Fabricate and/or install brake lines (double flare and ISO types); replace hoses, fittings, and supports as needed.	Priority Rating 2
B.8.	Select, handle, store, and fill brake fluids to proper level.	Priority Rating 1

B.9. Inspect, test, and/or replace metering (hold-off), proportioning (balance), pressure differential, and combination valves. Priority Rating 3
B.10. Inspect, test, replace and adjust height (load) sensing proportioning valve. Priority Rating 3
B.11. Inspect, test, and/or replace components of brake warning light system. Priority Rating 3
B.12. Bleed (manual, pressure, vacuum or surge) brake system. Priority Rating 1
B.13. Flush hydraulic system. Priority Rating 3

C. Drum Brake Diagnosis and Repair

C.1. Diagnose poor stopping, noise, pulling, grabbing, dragging or pedal pulsation concerns; determine necessary action. Priority Rating 1
C.2. Remove, clean (using proper safety procedures), inspect, and measure brake drums; determine necessary action. Priority Rating 1
C.3. Refinish brake drum. Priority Rating 2
C.4. Remove, clean, and inspect brake shoes, springs, pins, clips, levers, adjusters/self-adjusters, other related brake hardware, and backing support plates; lubricate and reassemble. Priority Rating 2
C.5. Remove, inspect, and install wheel cylinders. Priority Rating 2
C.6. Pre-adjust brake shoes and parking brake before installing brake drums or drum/hub assemblies and wheel bearings. Priority Rating 1
C.7. Install wheel, torque lug nuts, and make final checks and adjustments. Priority Rating 1

D. Disc Brake Diagnosis and Repair

D.1. Diagnose poor stopping, noise, pulling, grabbing, dragging, or pedal pulsation concerns; determine necessary action. Priority Rating 1
D.2. Remove caliper assembly from mountings; clean and inspect for leaks and damage to caliper housing; determine necessary action. Priority Rating 1
D.3. Clean and inspect caliper mounting and slides for wear and damage; determine necessary action. Priority Rating 1
D.4. Remove, clean, and inspect pads and retaining hardware; determine necessary action. Priority Rating 1
D.5. Disassemble and clean caliper assembly; inspect parts for wear, rust, scoring, and damage; replace seal, boot, and damaged or worn parts. Priority Rating 1
D.6. Reassemble, lubricate, and reinstall caliper, pads, and related hardware; seat pads, and inspect for leaks. Priority Rating 1
D.7. Clean, inspect, and measure rotor with a dial indicator and a micrometer; follow manufacturer's recommendations in determining need to machine or replace. Priority Rating 1
D.8. Remove and reinstall rotor. Priority Rating 2
D.9. Refinish rotor according to manufacturer's recommendations. Priority Rating 1
D.10. Adjust calipers with integrated parking brake system. Priority Rating 3
D.11. Install wheel, torque lug nuts, and make final checks and adjustments. Priority Rating 1

E. Power Assist Units Diagnosis and Repair

E.1. Test pedal free travel with and without engine running; check power assist operation. Priority Rating 2
E.2. Check vacuum supply (manifold or auxiliary pump) to vacuum-type power booster. Priority Rating 2
E.3. Inspect the vacuum-type power booster unit for vacuum leaks; inspect the check valve for proper operation; determine necessary action. Priority Rating 2

E.4. Inspect and test hydro-boost system and accumulator for leaks and proper operation; determine necessary action. Priority Rating 3

F. Miscellaneous (Wheel bearings, Parking Brakes, Electrical, etc.) Diagnosis and Repair

F.1. Diagnose wheel bearing noises, wheel shimmy, and vibration concerns; determine necessary action. Priority Rating 1

F.2. Remove, clean, inspect, repack, and install wheel bearings and replace seals; install hub and adjust wheel bearings. Priority Rating 1

F.3. Check parking brake cables and components for wear, rusting, binding, and corrosion; clean, lubricate, and replace as needed. Priority Rating 2

F.4. Check parking brake operation; determine necessary action. Priority Rating 1

F.5. Check operation of parking brake indicator light system. Priority Rating 3

F.6. Check operation of brake stop light system; determine necessary action. Priority Rating 1

F.7. Replace wheel bearing and race. Priority Rating 1

F.8. Inspect and replace wheel studs. Priority Rating 1

F.9. Remove and reinstall sealed wheel bearing assembly. Priority Rating 1

G. Anti-lock Brake System

G.1. Identify and inspect anti-lock brake system (ABS) components; determine necessary action. Priority Rating 2

G.2. Diagnose poor stopping, wheel lock-up, abnormal pedal feel or pulsation, and noise concerns caused by the anti-lock brake system (ABS); determine necessary action. Priority Rating 2

G.3. Diagnose anti-lock brake system (ABS) electronic control(s) and components using self-diagnosis and/or recommended test equipment; determine necessary action. Priority Rating 1

G.4. Depressurize high-pressure components of the anti-lock brake system (ABS). Priority Rating 2

G.5. Bleed the anti-lock brake system's (ABS) front and rear hydraulic circuits. Priority Rating 2

G.6. Remove and install anti-lock brake system (ABS) electrical/electronic and hydraulic components. Priority Rating 3

G.7. Test, diagnose, and service ABS speed sensors, toothed ring (tone wheel), and circuits using a graphing voltmeter (GMM), digital storage oscilloscope (DSO) (includes output signal, resistance, shorts to voltage/ground, and frequency data). Priority Rating 2

G.8. Diagnose anti-lock brake system (ABS) braking concerns caused by vehicle modifications (tire size, curb height, final drive ratio, etc.). Priority Rating 3

G.9. Identify traction control system components. Priority Rating 1

DEFINITION OF TERMS USED IN THE TASK LIST

To clarify the intent of these tasks, NATEF has defined some of the terms used in the task listings. To get a good understanding of what the task includes, refer to this glossary while reading the task list.

adjust To bring components to specified operational settings.

align To bring to precise alignment or relative position of components.

analyze To examine the relationship of components of an operation.

assemble (reassemble) To fit together the components of a device.

bleed To remove air from a closed system.

check To verify condition by performing an operational or comparative examination.

clean	To rid components of extraneous matter for the purpose of reconditioning, repairing, measuring and reassembling.
determine	To establish the procedure to be used to effect the necessary repair.
determine necessary action	Indicates that the diagnostic routine(s) is the primary emphasis of a task. The student is required to perform the diagnostic steps and communicate the diagnostic outcomes and corrective actions required addressing the concern or problem. The training program determines the communication method (worksheet, test, verbal communication, or other means deemed appropriate) and whether the corrective procedures for these tasks are actually performed.
diagnose	To locate the root cause or nature of a problem by using the specified procedure.
disassemble	To separate a component's parts as a preparation for cleaning, inspection, or service.
fill (refill)	To bring fluid level to specified point or volume.
flush	To use fluid to clean an internal system.
identify	To establish the identity of a vehicle or component prior to service; to determine the nature or degree of a problem.
inspect	(see *check*)
install (reinstall)	To place a component in its proper position in a system.
lubricate	To employ the correct procedures and materials in performing the prescribed service.
measure	To compare existing dimensions to specified dimensions by the use of calibrated instruments and gauges.
mount	To attach or place tool or component in proper position.
perform	To accomplish a procedure in accordance with established methods and standards.
perform necessary action	Indicates that the student is to perform the diagnostic routine(s) and perform the corrective action item. Where various scenarios (conditions or situations) are presented in a single task, at least one of the scenarios must be accomplished.
reassemble	(see *assemble*)
remove	To disconnect and separate a component from a system.
replace	To exchange an unserviceable component with a new or rebuilt component; to reinstall a component.
select	To choose the correct part or setting during assembly or adjustment.
service	To perform a specified procedure when called for in the owner's or service manual.
test	To verify condition through the use of meters, gauges, or instruments.
torque	To tighten a fastener to specified degree of tightness (in a given order or pattern if multiple fasteners are involved on a single component).
vacuum test	To determine the integrity and operation of a vacuum (negative pressure) operated component and/or system.

BRAKE SYSTEMS TOOLS AND EQUIPMENT

Many different tools and many kinds of testing and measuring equipment are used to service brake systems. NATEF has identified many of these and has said that a Brake technician must know what they are and how and when to use them. The tools and equipment listed by NATEF are covered in the following discussion. Also included are the tools and equipment you will use while completing the job sheets. Although you need to be more than familiar with and will be using common hand tools, they are not part of this discussion. You should already know what they are and how to use and care for them.

Machinist's Rule

A machinist's rule is very much like an ordinary ruler. Each edge of this measuring tool is divided into increments based on a different scale. A typical machinist's rule based on the USCS system of measurement may have scales based on 1/8-, 1/16-, 1/32-, and 1/64-inch intervals. Of course, metric machinist's rules are also available. Metric rules are usually divided into 0.5-mm and 1-mm increments.

Some machinist's rules are based on decimal intervals. These are typically divided into 1/10-, 1/50-, and 1/1,000-inch (0.1, 0.01, and 0.001) increments. Decimal machinist's rules are very helpful when measuring dimensions that are specified in decimals because you won't need to convert fractions to decimals.

Micrometers

A micrometer is used to measure linear outside and inside dimensions. Both outside and inside micrometers are calibrated and read in the same manner. Special brake rotor micrometers are available to measure rotor thickness and to check for parallelism. These typically start at a reading of 0.30-inches and extend to 1.30-inches. This range allows the micrometer to be used on most rotors. The major components and markings of a micrometer include the frame, anvil, spindle, locknut, sleeve, sleeve numbers, sleeve long line, thimble marks, thimble, and ratchet. Micrometers are calibrated in either inch or metric graduations and are available in a range of sizes.

To use and read a micrometer, choose the appropriate size for the object being measured. Typically micrometers measure an inch, therefore the range covered by one size of micrometer would be from 0 to 1 inch and another would measure 1 to 2 inches, and so on.

Open the jaws of the micrometer and slip the object between the spindle and anvil. While holding the object against the anvil, turn the thimble using your thumb and forefinger until the spindle contacts the object. Never clamp the micrometer tightly. Use only enough pressure on the thimble to allow the work to just fit between the anvil and spindle. To get accurate readings, you should slip the micrometer back and forth over the object until you feel a very light resistance, while at the same time rocking the tool from side to side to make certain the spindle cannot be closed any further. When a satisfactory adjustment has been made, lock the micrometer. Read the measurement scale.

The graduations on the sleeve each represent 0.025 inch. To read a measurement on a micrometer, begin by counting the visible lines on the sleeve and multiplying them by 0.025. The graduations on the thimble assembly define the area between the lines on the sleeve. The number indicated on the thimble is added to the measurement shown on the sleeve. This sum is the dimension of the object.

Micrometers are available to measure in 0.0001 (ten-thousandths) of an inch. Use this type of micrometer if the specifications call for that much accuracy.

A metric micrometer is read in the same way, except the graduations are expressed in the metric system of measurement. Each number on the sleeve represents 5 millimeters (mm) or 0.005 meter (m). Each of the ten equal spaces between each number, with index lines alternating above and below the horizontal line, represents 0.5 mm or five tenths of an mm. Therefore, one revolution of the thimble changes the reading one space on the sleeve scale or 0.5 mm. The beveled edge of the thimble is divided into 50 equal divisions with every fifth line numbered: 0, 5, 10, . . ., 45. Since one complete revolution of the thimble advances the spindle 0.5 mm, each graduation on the thimble is equal to one hundredth of a millimeter. As with the inch-graduated micrometer, the separate readings are added together to obtain the total reading.

Some technicians use a digital micrometer, which is easier to read. These tools do not have the various scales; rather the measurement is displayed and read directly off the micrometer.

Inside micrometers can be used to measure the inside diameter of a bore. To do this, place the tool inside the bore and extend the measuring surfaces until each end touches the bore's surface. If the bore is large, it might be necessary to use an extension rod to increase the micrometer's range. These extension rods come in various lengths. The inside micrometer is read in the same manner as an outside micrometer.

A depth micrometer is used to measure the distance between two parallel surfaces. A common use for a depth micrometer is the measurement of brake shoe lining thickness (Figure 2). The sleeves,

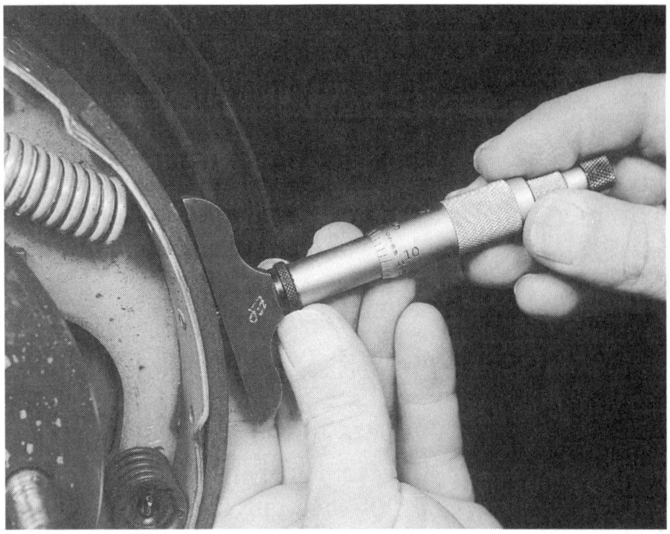

Figure 2 Measuring brake shoe thickness with a depth micrometer.

thimbles, and ratchet screws operate in the same way as other micrometers. Likewise, depth micrometers are read in the same way as other micrometers. A depth gauge is used to check the lining thickness of drum brake shoes.

Drum Micrometer

A drum micrometer (Figure 3) has one purpose, to measure the inside diameter of a brake drum. A drum micrometer has two movable arms; one arm has a dial indicator and the other has an outside anvil that fits against the inside of the drum. When using this micrometer to measure a drum, the arms are held to a shaft by lock screws that tighten into grooves in the shaft. The grooves are spaced one eighth of an inch apart on the shaft. Metric drum micrometers are similar except that the grooves are spaced at two-millimeter intervals. These features allow the micrometer to be used on nearly any size drum.

To use a drum micrometer, loosen the lock screws and move the arms so that the micrometer fits inside the drum. Then extend the arms until they are expanded to the nominal size of the drum. Place the micrometer inside the drum and rock it gently on the drum's surface until the highest reading on the indicator is obtained. It is best to take measurements across several planes of the drum. Doing this checks the drum for out-of-roundness.

Vernier Calipers

Vernier calipers can be used to measure inside, outside, depth, and height dimensions. The most common use of a vernier caliper when working on brakes is measuring brake pad thickness.

Dial Indicator

The dial indicator is calibrated in 0.001-inch (one-thousandth inch) increments. Metric dial indicators are also available. Both types are used to measure movement. Common uses of the dial indicator include measuring rotor (Figure 4) and axle flange runout. Dial indicators have many different attaching devices that can be used to connect the indicator to the component to be measured.

To use a dial indicator, position the indicator rod against the object to be measured. Then, push the indicator toward the work until the indicator needle travels far enough around the gauge face to permit movement to be read in either direction. Zero the indicator needle on the gauge. Move the object in the direction required, while observing the needle of the gauge. Always be sure the range of the dial indicator is sufficient to allow the amount of movement required by the measuring procedure. For example, never use a 1-inch indicator on a component that will move 2 inches.

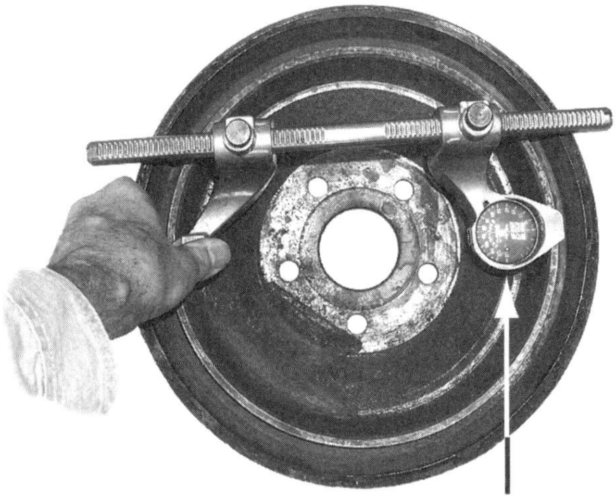

Figure 3 A brake drum micrometer.

Figure 4 Checking a rotor with a dial indicator.

Feeler Gauges

A feeler gauge is a precisely machined thin strip of metal. A feeler gauge set is a collection of these strips, each with a different and specific thickness. Feeler gauges are used to check piston-to-wheel cylinder bores and drum-to-brake shoe clearances.

Brake Shoe Adjusting Gauge (Calipers)

A brake shoe adjusting gauge (Figure 5) is an inside-outside measuring instrument. One half of the gauge is placed inside the brake drum and expanded to fit the diameter of the drum. Then, a lock screw is tightened to hold that dimension and the other side of the gauge is fit over the installed brake shoes. The brake shoes are then adjusted until the gauge slips over them. This provides a rough adjustment of the brake shoes; a final adjustment must be made after the drum is installed.

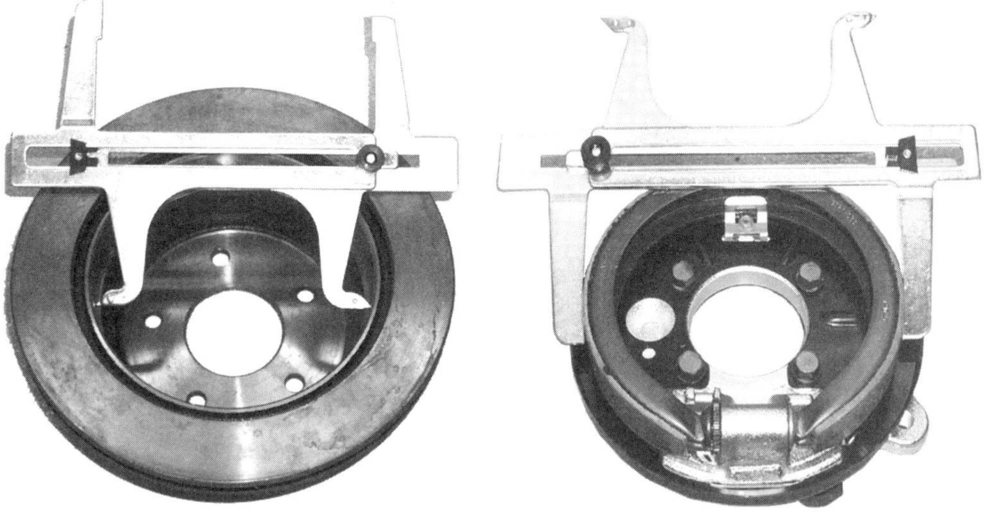

Figure 5 A drum brake adjusting gauge.

Brake Pedal Effort Gauge

A brake pedal effort gauge is used to diagnose power brake systems and to check brake pedal travel and effort. The tool is attached to the brake pedal and has a plunger and gauge that measures the force applied to the brake pedal.

Power Steering Pressure Gauge

A power steering pressure gauge is used to test the power steering pump pressure. This test is important when checking hydraulic boost brake systems. Since the power steering pump delivers extremely high pressure during this test, the recommended procedure in the vehicle manufacturer's service manual must be followed. However, the typical procedure for using a pressure gauge on power steering systems is given here as an example.

To check the pressure of the pump, a pressure gauge with a shut-off valve is needed. With the engine off, disconnect the pressure hose at the pump. Install the pressure gauge between the pump and the steering gear. Use any adapters that may be necessary to make good connections with the vehicle's system. Open the shut-off valve and bleed the system as described in the service manual.

Start the engine and run it for approximately two minutes or until the engine reaches normal operating temperature. Then stop the engine and add fluid to the power steering pump if necessary. Now restart the engine and allow it to idle. Observe the pressure reading. The readings should be about 30 to 80 psi (200 to 550 kPa). If the pressure is lower than what is specified, the pump may be faulty. If the pressure is greater than specifications, the problem may be restricted hoses.

Now close the shut-off valve, observe the pressure reading, and reopen the valve. Do not keep the valve closed for more than 5 seconds. With the valve closed the pressure should have increased to 600 to 1,300 psi (4,100 to 8,950 kPa). Check the pressure reading on the gauge; if the pressure is too high, a faulty pressure relief valve is suggested. If the pressure is too low, the pump may be bad.

Belt Tension Gauge

A belt tension gauge is used to measure drive belt tension. The belt tension gauge is installed over the belt, and the gauge indicates the amount of belt tension.

Multimeters

A multimeter is a must for diagnosing electrical and electronic systems. Multimeters have different names, depending on what they measure and how they function. A volt-ohm-milliamp meter is referred to as a VOM or DVOM, if it is digital. A DMM is a digital multimeter that can measure many more things than volts, ohms, and low current.

Most multimeters measure direct current (dc) and alternating current (ac) amperes, volts, and ohms. More advanced multimeters may also measure diode continuity, frequency, temperature, engine speed, and dwell and/or duty cycle.

Multimeters are available with either digital or analog displays. DMMs provide great accuracy by measuring volts, ohms, or amperes in tenths, hundredths, or thousandths of a unit. Several test ranges are usually provided for each of these functions. Some meters have multiple test ranges that must be manually selected; others are auto-ranging.

Analog meters use a sweeping needle against a scale to display readings and are not as precise as digital meters. Analog meters have low input impedance and should not be used on sensitive electronic circuits or components. Digital meters have high impedance and can be used on electronic circuits, as well as electrical circuits.

Scan Tools

The introduction of computer-controlled systems brought with it the need for tools capable of troubleshooting electronic control systems. There are a variety of computer scan tools available today that do just that. A scan tool is a microprocessor designed to communicate with the vehicle's computer. Connected to the computer through diagnostic connectors, a scan tool can access trouble codes, run tests to check system operations, and monitor the activity of the system. Trouble codes and test results are displayed on an LED screen, or printed out on the scanner printer.

Scan tools retrieve fault codes from a computer's memory and digitally display these codes on the tool. Many scan tools also can activate system functions to test individual components. A scan tool may also perform many other diagnostic functions depending on the year and make of the vehicle. Most aftermarket scan tools have removable modules that are updated each year. These modules are designed to test the computer systems on various makes of vehicles. For example, some scan testers have a 3-in-1 module that tests the computer systems on Chrysler, Ford, and General Motors vehicles. A 10-in-1 module is also available to diagnose computer systems on vehicles imported by 10 different manufacturers. These modules plug into the scan tool.

Scan tools are capable of testing many onboard computer systems, such as transmission controls, engine computers, antilock brake computers (Figure 6), air bag computers, and suspension computers, depending on the year and make of the vehicle and the type of scan tester. In many cases, the technician must select the computer system to be tested with the scanner after it has been connected to the vehicle.

The scan tool is connected to specific diagnostic connectors on various vehicles. Most manufacturers have one diagnostic connector that connects the data wire from each onboard computer to a specific terminal in the connector. Other vehicle manufacturers have several different diagnostic connectors on each vehicle, and each of these connectors may be connected to one or more onboard computers. A set of connectors is supplied with the scanner to allow tester connection to various diagnostic connectors on different vehicles.

The scanner must be programmed for the model year, make of vehicle, and type of engine. With some scan tools, this selection is made by pressing the appropriate buttons on the tester, as directed by the digital tester display. On other scan testers, the appropriate memory card must be installed in the tester for the vehicle being tested. Some scan testers have a built-in printer to print test results, while other scan testers may be connected to an external printer.

As automotive computer systems become more complex, the diagnostic capabilities of scan testers continue to expand. Many scan testers now have the capability to store, or "freeze," data into the tester during a road test, and then play back the data when the vehicle is returned to the shop.

Some scan testers now display diagnostic information based on the fault code in the computer memory. Service bulletins published by the manufacturer of the scan tester may be indexed by the tester after the vehicle information is entered in the tester. Other scan testers display sensor specifications for the vehicle being tested.

Trouble codes are only set by the vehicle's computer when a voltage signal is entirely out of its normal range. The codes help technicians identify the cause of the problem when this is the case. If a signal is within its normal range but is still not correct, the vehicle's computer will not display a trouble code. However, a problem will still exist. As an aid to identify this type of problem, most manufacturers recommend that the signals to and from the computer be carefully looked at. This is done

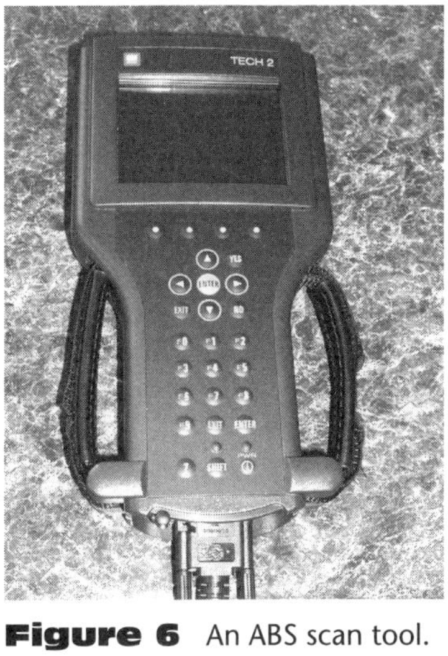

Figure 6 An ABS scan tool.

through the use of a scan tool or breakout box. A breakout box allows the technician to check voltage and resistance readings between specific points within the computer's wiring harness.

With OBD-II, the diagnostic connectors are located in the same place on all vehicles. Also, any scan tools designed for OBD-II will work on all OBD-II systems, therefore the need to have designated scan tools or cartridges is eliminated. The OBD-II scan tool has the ability to run diagnostic tests on all systems and has "freeze frame" capabilities.

Floor Jack

A floor jack is a portable unit mounted on wheels. The lifting pad on the jack is placed under the chassis of the vehicle, and the jack handle is operated with a pumping action. This forces fluid into a hydraulic cylinder in the jack, and the cylinder extends to force the jack lift pad upward and lift the vehicle. Always be sure that the lift pad is positioned securely under one of the car manufacturer's recommended lifting points. To release the hydraulic pressure and lower the vehicle, the handle or release lever must be turned slowly.

The maximum lifting capacity of the floor jack is usually written on the jack decal. Never lift a vehicle that exceeds the jack lifting capacity. This action may cause the jack to break or collapse, resulting in vehicle damage or personal injury.

Safety Stands

Whenever a vehicle is raised by a jack, it must be supported by safety (jack) stands. Check the service manual for the proper locations for positioning the jack and the safety stands.

Lift

A lift is used to raise a vehicle so the technician can work under the vehicle. The lift arms must be placed under the car manufacturer's recommended lifting points prior to raising a vehicle. Twin posts are used on some lifts, whereas other lifts have a single post. Some lifts have an electric motor, which drives a hydraulic pump to create fluid pressure and force the lift upward. Other lifts use air pressure from the shop air supply to force the lift upward. If shop air pressure is used for this purpose, the air pressure is applied to fluid in the lift cylinder. A control lever or switch is placed near the lift. The control lever supplies shop air pressure to the lift cylinder, and the switch turns on the lift pump motor. Always be sure that the safety lock is engaged after the lift is raised. When the safety lock is released, a release lever is operated slowly to lower the vehicle.

Hydraulic Press

When two components have a tight precision fit between them, a hydraulic press is used to either separate these components or press them together. The hydraulic press rests on the shop floor, and an adjustable steel beam bed is held to the lower press frame with heavy steel pins. A hydraulic cylinder and ram are mounted on the top part of the press with the ram facing downward toward the press bed. The component being pressed is placed on the press bed with appropriate steel supports. A hand-operated hydraulic pump is mounted on the side of the press. When the handle is pumped, hydraulic fluid is forced into the cylinder, and the ram is extended against the component on the press bed to complete the pressing operation. A pressure gauge on the press indicates the pressure applied from the hand pump to the cylinder. The press frame is designed for a certain maximum pressure, and this pressure must not be exceeded during hand pump operation.

Hand-Held Grease Gun

A hand-operated grease gun forces grease into a grease fitting. Often these are preferred because the pressure of the grease can be controlled by the technician. However, many shops use low air pressure to activate a pneumatic grease gun. The suspension and steering system may have several grease or zerk fittings.

Torque-Indicating Wrench

Torque is the twisting force used to turn a fastener against the friction between the threads and between the head of the fastener and the surface of the component. The fact that practically every vehicle and engine manufacturer publishes a list of torque recommendations is ample proof of the importance of using proper amounts of torque when tightening nuts or bolts. The amount of torque applied to a fastener is measured with a torque-indicating or torque wrench.

There are three basic types of torque-indicating wrenches available with pounds per inch and pounds per foot increments: a beam torque wrench that has a beam that points to the torque reading, a "click"-type torque wrench in which the desired torque reading is set on the handle (when the torque reaches that level, the wrench clicks), and a dial torque wrench that has a dial that indicates the torque exerted on the wrench. Some designs of this type torque wrench have a light or buzzer that turns on when the desired torque is reached.

Flare Nut (Line) and Bleeder Screw Wrenches

Flare nut wrenches should be used to loosen and tighten brake line or tubing fittings. Flare nut wrenches surround the nut and provide a better grip on the fitting. They have a section cut out so that the wrench can be slipped around the brake line and dropped over the flare nut.

Special bleeder screw wrenches should be used to open and close bleeder screws. Bleeder screw wrenches are small, six-point box wrenches with offset handles that allow for easy access to the bleeder screws. Never use an open-end wrench on a bleeder screw. An open-end wrench can easily slip and round off the head of the bleeder screw.

Gear and Bearing Pullers

Many tools are designed for a specific purpose. An example of a special tool is a gear and bearing puller. Many gears and bearings have a slight interference fit (press-fit) when they are installed on a shaft or in a housing. Something that has a press-fit has an interference fit. For example, if the inside diameter of a bore is 0.001 inch smaller than the outside diameter of a shaft, when the shaft is fitted into the bore it must be pressed in to overcome the 0.001 inch interference fit. This press-fit prevents the parts from moving on each other. The removal of these gears and bearings must be done carefully to prevent damage to the gears, bearings, or shafts. Prying or hammering can break or bind the parts. A puller with the proper jaws and adapters should be used to remove gears and bearings. Using the proper puller, the force required to remove a gear or bearing can be applied with a slight and steady motion.

Special tools are also required to drive in bearing races (Figure 7). These tools must fit the outside diameter of the race without contacting the bore the races are being installed in.

Figure 7 Bearing races are best installed with the correct driver.

Bushing and Seal Pullers and Drivers

Another commonly used group of special tools are the various designs of bushing and seal drivers and pullers. Pullers are either a threaded or slide hammer type tool. Always make sure you use the correct tool for the job; bushings and seals are easily damaged if the wrong tool or procedure is used. Car manufacturers and specialty tool companies work closely together to design and manufacture special tools required to repair cars. Most of these special tools are listed in the appropriate service manuals.

Seal drivers are designed to fit squarely against the seal case and inside the seal lip. A soft hammer is used to tap the seal driver and drive the seal straight into the housing. Some tool manufacturers market a seal driver kit with drivers to fit many common seals.

Tie-Rod End and Ball Joint Puller

Some car manufacturers recommend a tie-rod end and ball joint puller to remove tie-rod ends and pull ball joint studs from the steering knuckle. A tie rod end remover is a safer and easier way of separating ball joints than a pickle fork.

Ball joint removal and pressing tools are designed to remove and replace pressed-in ball joints on front suspension systems. Often these tools are used in conjunction with a hydraulic press. The size of the removal and pressing tool must match the size of the ball joint.

Some ball joints are riveted to the control arm, and the rivets are drilled out for removal.

Front Bearing Hub Tool

Front bearing hub tools are designed to remove and install front wheel bearings on front-wheel-drive cars. These bearing hub tools are usually designed for a specific make of vehicle and the correct tools must be used for each application. Failure to do so may result in damage to the steering knuckle or hub. And the use of the wrong tool will waste quite a bit of your time.

Boot Drivers, Rings, and Pliers

Caliper dust boots are positioned between the caliper cylinder and the pistons to keep dirt and moisture out of the caliper bores. A special driver is used to properly install the dust boot. The driver is centered on the boot and the caliper bore and the boot is driven into the bore with a hammer. Some dust boots are not driven into the cylinder bore; instead, they are fitted into a special groove at the top of the cylinder bore. Special tools or pliers are used to expand the opening in the dust boot just enough to let the piston slide into the cylinder bore through the boot.

Caliper Piston Removal Tools

If it is difficult to pull or turn the caliper piston from the cylinder bore, pliers designed to remove pistons will make the job easier. These pliers grip the inside of the piston so you pull the piston out. There are also tools that use hydraulic pressure to force the piston out of its bore.

Hold-down and Return Spring Tools

Brake shoe return springs require special tools (Figure 8) for removal and installation. Most return spring tools are much like pliers fitted with special sockets and hooks to release and install the spring ends. A hold-down spring tool is much like a nut driver with a specially shaped end that fits over the spring retainer and allows the retainer to be rotated.

Drum Brake Adjusting Tools

Brake shoes must be adjusted after they are installed. Brake adjusting tools (spoons) and wire hooks are designed to allow easy access to the brake's star wheel adjusters so they can be turned and the brakes adjusted.

Tubing Tools

Brake lines are often purchased preformed and of the correct length to fit specific locations on specific vehicles. Straight brake lines can also be purchased in many lengths and several diameters. These must be bent to fit around other components. At times it may be necessary to fabricate the brake line. To do this you will need to cut, bend, and form flared ends on the lines. To do this you will need a tubing cutter and reamer, tube benders, a double flaring tool for SAE flares, and an ISO flaring tool.

Brake Cylinder Hones

Cylinder hones are used to clean corrosion, surface mars, and dirt from the bores of master cylinders, wheel cylinders, and calipers. Most cylinder hones have two or three replaceable abrasive stones at the ends of spring-loaded arms. To maintain proper stone pressure against the cylinder walls, the spring tension on the stones is typically adjustable. The shaft of the hone is mounted in a drill motor. Brush hones are often used. These have abrasive balls attached to flexible metal arms mounted to the hone's shaft. As the drill motor rotates the hone, centrifugal force moves the balls against the cylinder walls.

During the honing process, it is important to keep the honing stones well lubricated with brake fluid. When you have finished honing the cylinder, flush it thoroughly with denatured alcohol to remove all abrasives and dirt.

Brake Lathes

Brake lathes (Figure 9) are used to resurface brake rotors and drums. This typically involves cutting away very small amounts of metal to restore the surface of the rotor or drum. There are two basic types of lathes: a bench lathe and an on-the-car lathe. Both work in a similar fashion, but the on-the-car lathe only works on brake rotors. The rotor does not need to be removed to use this type of lathe. The bench lathe can resurface rotors and drums, but the rotor or drum must be removed from the vehicle first.

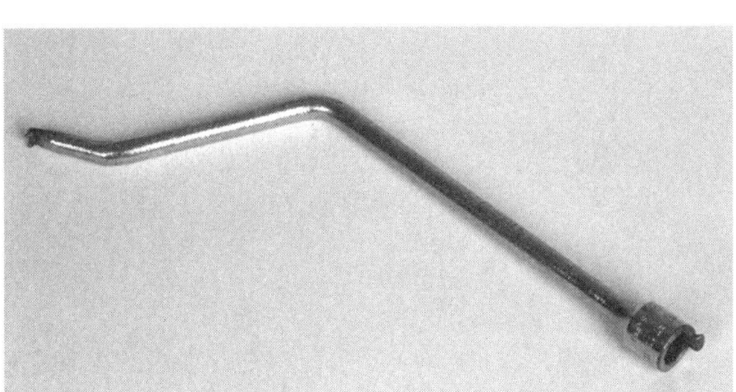

Figure 8 A brake spring tool.

Figure 9 A brake lathe. *Courtesy of Hennessy Industries, Inc.*

A cutting bit passes over the friction surface of the rotating rotor or drum to remove a small amount of metal. Most lathes include attachments for applying a final surface finish to the rotor or for grinding hard spots on drums.

Pressure Bleeders

Removing the air from the closed hydraulic brake system is very important. This is done by bleeding the system. Bleeding can be done manually, with a vacuum pump, or with a pressure bleeder (Figure 10). The latter is preferred because it is quick and very efficient.

A pressure bleeder consists of an adapter that fits over the brake fluid reservoir in place of the reservoir cap, hoses, and a tank. The tank is separated into two sections by a diaphragm. The top section of the tank is filled with brake fluid and the bottom part is exposed to compressed air. When the shop air enters the bottom section, it pushes on the diaphragm, which puts pressure on the brake fluid in the top section.

A hose connects the top of the tank to the master cylinder. The pressurized brake fluid from the tank flows into the master cylinder and out through the brake bleeder screws, forcing all air out of the lines.

Brake bleeding may involve other special tools, such as a syringe, master cylinder bleeder tubes, and assorted line and port plugs.

Negative-Pressure Enclosure and HEPA Vacuum Systems

Negative-pressure enclosures are designed to prevent dirt and asbestos fibers from entering the shop's air during brake work. Using these enclosures, brake system cleaning and inspection are performed inside a tightly sealed protective enclosure that covers and contains the brake assembly. The enclosure assembly is clear so work can safely and efficiently be done and has built-in gloves that allow you to work without actually touching the parts.

A HEPA (high efficiency particulate air) filter vacuum keeps the enclosure under negative pressure. Because particles cannot escape the enclosure, compressed air can be used to remove dust, dirt, and potential asbestos fibers from brake parts. The vacuum can also be used to draw out any asbestos-containing residue from brake parts.

HEPA vacuum cleaners can be used without the enclosure to clean shop areas and to clean brake parts. When working on a brake system, make sure the vacuum cleaner has a HEPA filter. After vacuuming, wipe the components with a damp cloth and discard the cloth according to the applicable hazardous waste regulation.

Figure 10 A pressure bleeder unit.

Low-Pressure Wet Cleaning Systems

A safe alternative to the negative-pressure enclosure system is the use of low-pressure wet cleaning systems. These are designed to wash dirt off brake parts and to catch the contaminated cleaning solution in a container. The cleaning solution can be used before and after the brake drum has been removed. Clean all parts before and during service.

Service Manuals

Perhaps the most important tools you will use are service manuals. There is no way a technician can remember all of the procedures and specifications needed to correctly repair all vehicles. Therefore, a good technician relies on service manuals and other information sources for this information. Good information plus knowledge allows a technician to fix a problem with the least bit of frustration and at the lowest cost to the customer.

Service manuals are needed to obtain specifications on torque values and critical measurements such as drum and rotor discard limits. To obtain the correct specifications and other information, you must first identify the vehicle you are working on. The best source for positive identification is the VIN.

The primary source of repair and specification information for any car, van, or truck is the manufacturer. The manufacturer publishes service manuals each year, for every vehicle built. Because of the enormous amount of information, some manufacturers publish more than one manual per year per car model. Manuals are typically divided into sections based on the major systems of the vehicle. Manufacturers' manuals cover all repairs, adjustments, specifications, detailed diagnostic procedures, and special tools required.

Since many technical changes occur on specific vehicles each year, manufacturers' service manuals need to be constantly updated. Updates are published as service bulletins (often referred to as Technical Service Bulletins or TSBs) that show the changes in specifications and repair procedures during the model year. These changes do not appear in the service manual until the next year. The car manufacturer provides these bulletins to dealers and repair facilities on a regular basis.

Service manuals are also published by independent companies rather than the manufacturers. However, they pay for and get most of their information from the car makers. These manuals contain component information, diagnostic steps, repair procedures, and specifications for several car makes in one book. Information is usually condensed and is more general in nature than the manufacturer's manuals. The condensed format allows for more coverage in less space and, therefore, is not always specific. They may also contain several years of models as well as several car makes in one book.

Many of the larger parts manufacturers have excellent guides on the various parts they manufacture or supply. They also provide updated service bulletins on their products. Other sources for up-to-date technical information are trade magazines and trade associations.

The same information that is available in service manuals and bulletins is also available on CD-ROMs and DVDs. A single compact disk can hold 250,000 pages of text, which eliminates the need for a huge library containing all the printed manuals. Using a CD-ROM to find information is also easier and quicker. The disks are normally updated monthly, and not only contain the most recent service bulletins but also engineering and field service fixes.

CROSS-REFERENCE GUIDE

NATEF Task	Job Sheet
A.1	38
A.2	39
A.3	39
B.1	40
B.2	1
B.3	2
B.4	3
B.5	4
B.6	5
B.7	5
B.8	6
B.9	7
B.10	8
B.11	9
B.12	10
B.13	6
C.1	11
C.2	12
C.3	13
C.4	12
C.5	14
C.6	12
C.7	15
D.1	16
D.2	17
D.3	17
D.4	17
D.5	18
D.6	18
D.7	19
D.8	23
D.9	20
D.10	21
D.11	22
E.1	24
E.2	25
E.3	25
E.4	26

NATEF Task	Job Sheet
F.1	27
F.2	28
F.3	29
F.4	30
F.5	9
F.6	9
F.7	28
F.8	41
F.9	42
G.1	31
G.2	31
G.3	32
G.4	33
G.5	34
G.6	35
G.7	36
G.8	37
G.9	43

JOB SHEETS

BRAKES JOB SHEET 1

Pedal Freeplay Inspection and Adjustment

Name ______________________ Station ______________ Date ______________

NATEF Correlation

This Job Sheet addresses the following NATEF task:

B.2. Measure brake pedal height; determine necessary action.

Objective

Upon completion of this job sheet, you will be able to correctly measure and adjust pedal height.

Tools and Materials

12-inch rule

Basic hand tools

Protective Clothing

Goggles or safety glasses with side shields

Describe the vehicle being worked on:

Year ________ Make ____________________ Model ____________________

VIN ____________________ Engine type and size ____________________

PROCEDURE

1. Pump the brake pedal with the engine off to exhaust vacuum in the booster. ☐ Task completed
2. Place a ruler against the car floor, in line with the arc of pedal travel. ☐ Task completed
3. Press the pedal by hand and measure the amount of travel before looseness in the linkage, or freeplay, is taken up. ☐ Task completed

4. Measure at the top or bottom of the pedal, whichever provides the most accurate view. Compare your measurement with the vehicle specifications. Summarize your results.

 ____

 ____

5. To adjust freeplay, the pushrod at the pedal must be lengthened or shortened. ☐ Task completed
6. To do this, loosen the locknut on the pushrod at the pedal and rotate the pushrod while rechecking freeplay measurement. ☐ Task completed
7. Tighten the locknut when adjustment is correct. ☐ Task completed
8. If the car has a mechanical brake lamp switch located on the brake pedal linkage, check switch operation and adjust it if necessary after adjusting pedal freeplay. ☐ Task completed

Instructor's Comments

____

____

____

BRAKES JOB SHEET 2

Checking a Master Cylinder for Leaks

Name ______________________________ Station __________________ Date __________________

NATEF Correlation

This Job Sheet addresses the following NATEF task:

B.3. Check master cylinder for internal and external leaks and proper operation; determine necessary action.

Objective

Upon completion of this job sheet, you will be able to check a master cylinder for internal and external leaks and for proper operation.

Tools and Materials

Basic hand tools

Protective Clothing

Goggles or safety glasses with side shields

Describe the vehicle being worked on:

Year __________ Make __________________________ Model __________________________

VIN __________________________ Engine type and size __________________________

PROCEDURE

1. Check the fluid level in the master cylinder reservoir and if necessary add fluid to correct the level. What type of fluid is recommended for this vehicle?

   ____

2. Carefully check the condition of the fluid. The fluid should be clear and transparent, although some darkening is acceptable. Describe your findings and state what is indicated by the condition of the fluid.

   ____

   ____

3. Check for unequal fluid levels in the master cylinder reservoir chambers on front disc and rear drum systems. Describe your findings.

   ____

4. Check for evidence of leaks or cracks in the master cylinder housing. Describe your findings.

   ____

5. If the master cylinder does not appear to be leaking, raise the vehicle on a lift and inspect all brake lines, hoses, and connections. Look for brake fluid on the floor under the vehicle and at the wheels. Describe your findings.

____

____

6. Check the brake lines for kinks, dents, or other damage. Also check for signs of leakage. Describe your findings.

____

____

7. Check the brake hoses. They should be flexible and free of leaks, cuts, cracks, and bulges. Describe your findings.

____

____

8. Inspect the backing plates for fluid and grease. Describe your findings. Also make sure all parts attached to the plates are securely fastened.

____

____

9. To determine whether or not the brake system has an external leak, run the engine at idle with the transmission in neutral. ☐ Task completed

10. Depress the brake pedal and hold it down with a constant foot pressure. The pedal should remain firm and the footpad should be at least 2 inches from the floor for manual brakes and 1 inch for power brakes without ABS. ☐ Task completed

11. Hold the pedal depressed with medium foot pressure for about 15 seconds to make sure that the pedal does not drop under steady pressure. If the pedal drops under steady pressure, the master cylinder or a brake line or hose may be leaking. ☐ Task completed

12. If there appears to be no external leak but the BRAKE warning lamp is lit, the master cylinder may be bypassing or losing pressure internally. ☐ Task completed

13. To check for an internal master cylinder leak, remove the master cylinder cover and be sure the reservoirs are at least half full. ☐ Task completed

14. Watch the fluid levels in the reservoirs while an assistant slowly presses the brake pedal and then quickly releases it. ☐ Task completed

15. If fluid level rises slightly under steady pressure, the piston cups are probably leaking. Fluid level rising in one reservoir and falling in the other as the brake pedal is pressed and is released also can indicate that fluid is bypassing the piston cups. Describe your findings.

____

____

____

Instructor's Comments

BRAKES JOB SHEET 3

Bench Bleed a Master Cylinder

Name ______________________________ Station ________________ Date ________________

NATEF Correlation

This Job Sheet addresses the following NATEF task:

B.4. Remove, bench bleed, and reinstall master cylinder.

Objective

Upon completion of this job sheet, you will be able to bench bleed a master cylinder.

Tools and Materials

Brake fluid
Clean rag
Service manual
Tubing
Vise
Wooden dowel or smooth, rounded rod
Wrench

Protective Clothing

Goggles or safety glasses with side shields

Describe the vehicle that the master cylinder came from:

Year __________ Make ________________________ Model ________________________

VIN ________________________ Engine type and size ________________________

PROCEDURE

1. Mount the cylinder in a vise. Do not apply excessive pressure to the casting. Ensure that the bore is horizontal. Remove the master cylinder cover. ☐ Task completed
2. Connect the short lengths of tubing to the outlet ports. Then bend them upward and place their open end in each reservoir. Fill the reservoirs with fresh brake fluid. ☐ Task completed
3. Use a wooden dowel or smooth, rounded rod to slowly pump the master cylinder until bubbles stop appearing at the tube ends. Make sure both pistons are completely bottomed in the bore. On release, allow the pistons to return completely to a stop. ☐ Task completed
4. Using the same tool, push the piston 1/4 to 1/2 inch from its fully released position. Allow the piston to return about 1/8 inch. Then push it back to the initial 1/4 to 1/2 inch position. Rapidly move the piston back and forth in this manner until the air bubbles stop coming from the ports located inside the reservoirs. ☐ Task completed
5. Remove the tubing. Refill the reservoirs. Secure the master cylinder cover. ☐ Task completed

6. Install the master cylinder on the vehicle. Attach the lines, but do not tighten the tube connections. ☐ Task completed

7. Loosen the nut, then slowly depress the brake pedal to force any trapped air out of the connections. Tighten the nut slightly before releasing the pedal. ☐ Task completed

8. Have all air bubbles been removed? If not, repeat the procedure. ☐ Task completed

9. Tighten the connections to the manufacturer's specifications. ☐ Task completed

10. Make sure the master cylinder reservoirs are adequately filled with brake fluid. ☐ Task completed

Problems Encountered

____

____

____

Instructor's Comments

____

____

____

BRAKES JOB SHEET 4

Brake System Road Test

Name ______________________ Station ______________ Date ______________

NATEF Correlation

This Job Sheet addresses the following NATEF task:

B.5. Diagnose poor stopping, pulling or dragging concerns caused by malfunctions in the hydraulic system; determine necessary action.

Objective

Upon completion of this job sheet, you will be able to diagnose poor stopping, pulling or dragging problems and check for normal operation of the brake system.

Protective Clothing

Goggles or safety glasses with side shields

Describe the vehicle being worked on:

Year ________ Make ____________________ Model ____________________

VIN ____________________ Engine type and size ____________________

PROCEDURE

1. Inspect the vehicle for worn, mismatched, or underinflated or overinflated tires. What did you find?

   ____

   ____

2. Inspect the vehicle for unequal vehicle loading. A heavily loaded vehicle requires more braking power. If the load is unequal from front to back or side to side, the brakes may grab or pull to one side. What did you find?

   ____

   ____

3. Inspect the vehicle for evidence of wheel misalignment. Check the tire wear patterns. What did you find?

   ____

   ____

4. If the tires are in good shape and the wheel alignment and vehicle loading do not appear to be the problem, proceed with a brake system road test. Test drive the vehicle on a dry, clean, relatively smooth roadway or parking lot. ☐ Task completed

5. Test the vehicle at low speeds. Use both light and fairly heavy pedal pressure. If the system can safely handle it, test the vehicle at higher speeds. Avoid locking the brakes and skidding the tires. If the vehicle reacts abnormally or makes noise during braking, write out a description of what it is doing and when it does it. Use these questions as guidelines for evaluating brake performance.
   a. Do you hear squeals or grinding?
   b. Do the brakes grab or pull to one side?
   c. Does the brake pedal feel spongy or hard when applied?
   d. Do the brakes release promptly when you take your foot off the brake pedal?

____

____

____

6. Check the brake warning lamp on the instrument panel. It should light when the ignition switch is in the start position and go off when the ignition returns to the run position. Did this happen?

____

____

7. If the brake warning lamp stays on when the ignition is on, make sure the parking brake is fully released. If it is, the problem may be a low brake fluid level in the master cylinder. Some vehicles have a separate master cylinder fluid level warning lamp. If either warning lamp remains on, check the fluid level in the master cylinder reservoir. What did you find?

____

____

8. Summarize the behavior of the vehicle's brakes and list the probable causes for any abnormal performance and noise.

____

____

____

____

Instructor's Comments

____

____

____

BRAKES JOB SHEET 5

Brake Line, Fitting, and Hose Service

Name ______________________ Station ______________ Date ______________

NATEF Correlation

This Job Sheet addresses the following NATEF tasks:

B.6. Inspect brake lines, flexible hoses, and fittings for leaks, dents, kinks, rust, cracks, bulging or wear; tighten loose fittings and supports; determine necessary action.

B.7. Fabricate and/or install brake lines (double flare and ISO types); replace hoses, fittings, and supports as needed.

Objective

Upon completion of this job sheet, you will be able to inspect brake lines, flexible hoses, and fittings; tighten loose fittings and supports; fabricate and install brake lines; and replace hoses, fittings, and supports.

Tools and Materials

Flare nut wrenches Double flare tool

Tubing cutter ISO flaring tool

Protective Clothing

Goggles or safety glasses with side shields

Describe the vehicle being worked on:

Year ________ Make ____________________ Model ____________________

VIN ____________________ Engine type and size ____________________

PROCEDURE

1. Steel brake lines should be checked for rust, corrosion, dents, kinks, and cracks. What did you find?

____

____

2. Check the line and hose mounting clips, brackets, and fittings where hoses are connected to rigid brake lines. What did you find?

____

____

3. Check for missing mounting clips. Inspect all brake tubing for looseness and look for empty screw holes or scuff marks on body and frame parts that indicate missing clips. What did you find?

____

____

4. Inspect brake hoses for abrasion caused by rubbing against chassis parts and for cracks at stress points, particularly near fittings. What did you find?

____

____

5. Look for fluid seepage indicated by softness in the hose, accompanied by a dark stain on the outer surface. What did you find?

____

____

6. Look at each hose closely for general damage and deterioration such as cracks, soft or spongy feel or appearance, stains, blisters, and abrasions. What did you find?

____

____

7. To check for internal hose damage, have an assistant pump the brakes and feel the hose for swelling or bulging as pressure is applied internally. Have your assistant apply and release the brakes; then quickly spin the wheel. If the brake at any wheel seems to drag after pressure is released, the brake hose may be restricted internally. What did you find?

____

____

8. Make sure the brake hoses are not kinked or twisted. While watching the front brakes, have your assistant rotate the steering wheel from lock to lock and verify that the brake hoses do not rub on chassis parts or twist and kink when the wheels turn. What did you find?

____

____

Brake Hose Replacement

1. Make sure the replacement brake hose is the same length as the original one. ☐ Task completed

2. Clean dirt away from the fittings at each end of the hose to keep it from entering the system. ☐ Task completed

3. Disconnect the swivel fitting first if the hose is equipped with one. If the hose has fixed fittings, disconnect the female end first. If the hose has a banjo fitting on one end, disconnect it first. Use a flare nut wrench to loosen and disconnect the fittings. When loosening one fitting at the end of a hose, hold the mating half of the fitting with another flare nut wrench. ☐ Task completed

4. Remove the hose retaining clip from the mounting bracket with a pair of pliers. ☐ Task completed

5. Separate the hose from the mounting bracket and any other clips used to hold it in place. ☐ Task completed

6. Use a flare nut wrench to disconnect the other end of the hose from the caliper or wheel cylinder. If the hose has a banjo fitting, use a box wrench to hold the banjo fitting while loosening the bolt. ☐ Task completed

7. If the hose has a fixed male end, install it into the wheel cylinder or caliper first. If the connection requires a copper gasket, install a new one. ☐ Task completed

8. If one end of the hose has a banjo fitting for attachment to a caliper, install the banjo bolt and a new copper gasket on each side of the fitting. Leave the banjo bolt loose at this time; tighten it after connecting and securing the other end of the hose. ☐ Task completed

9. Route the hose through any support devices and install any required locating clips. ☐ Task completed

10. Insert the free end of the hose through the mounting bracket. ☐ Task completed

11. Connect the flare nut on the steel brake line to the female end of the hose or connect the swivel end of the hose to the mating fitting. Use a flare nut wrench to tighten the fitting and hold the hose with another flare nut wrench to keep it from twisting. Check the colored stripe or the raised rib on the outside of the hose to verify that the hose has not twisted during installation. ☐ Task completed

12. Install the retaining clip and any other clips to hold the hose to its mounting bracket. ☐ Task completed

13. Position the banjo fitting to provide the best hose position and tighten the bolt. ☐ Task completed

14. After installing the new brake hose, check the hose and line connections for leaks and tighten if needed. Check for clearance during suspension rebound and while turning the wheels. If any contact occurs, reposition the hose, adjusting only the female end or the swivel end. ☐ Task completed

Brake Tubing Replacement

1. Clean the dirt away from the fittings at each end of the tubing. ☐ Task completed

2. Use a flare nut wrench to disconnect the fittings at each end of the tubing. If the tubing is attached to a hose, use another flare nut wrench to hold the hose fitting. ☐ Task completed

3. Remove the mounting clips from the chassis and remove the brake tubing. Inspect the clips and their screws to determine if they are reusable. ☐ Task completed

4. To install a length of brake tubing, position it on the chassis and install the mounting clips loosely. ☐ Task completed

5. Next, use the appropriate flare nut wrenches to connect the fittings at both ends of the tubing. ☐ Task completed

6. Tighten the fittings securely and then tighten the mounting clips. ☐ Task completed

Fabricating Brake Tubing

NOTE: *Always use prefabricated tubing whenever possible. Double-wall steel brake tubing is the only type of tubing approved for brake lines. Never use copper tubing or any other tubing material as a replacement; it cannot withstand the high pressure or the vibrations to which a brake line is exposed. Fluid leakage and system failure can result.*

1. To form a replacement length of tubing, determine the exact length of replacement tubing that is needed; add 1/8 inch for each flare that is to be made. ☐ Task completed
2. Hold the free outer end of the tubing against a flat surface with one hand and unroll the roll of tubing in a straight line with the other hand. Do not lay the roll flat and pull one end toward you; this will twist and kink the tubing. ☐ Task completed
3. Mark the tubing at the point to be cut and place a tubing cutter on the tubing. Tighten the cutter until the cutting wheel contacts the tubing at the marked point. Turn the cutter around the tubing toward the open side of the cutter jaws. After each revolution, tighten the cutter slightly until the cut is made. ☐ Task completed
4. Ream the cut end of the tubing with the reaming tool on the tubing cutter to remove burrs and sharp edges. Hold the end downward so that metal chips fall out. Ream only enough to remove burrs; then blow compressed air through the tubing to be sure all chips are removed. ☐ Task completed
5. With a tube bender, match the turns and curves of the old brake line. ☐ Task completed
6. After the tubing is cut and bent to fit, flares must be formed on any unfinished ends. ☐ Task completed
7. What type of flare is required for the brake line fittings?

   ____

8. Place the fittings onto the fabricated brake line before forming the flare. Make sure the flare nut threads are facing the end of the tube. ☐ Task completed
9. A double flare is made in two stages using a special flaring tool. Select the forming die from the flaring kit that matches the inside diameter of the tubing. ☐ Task completed
10. Clamp the tubing in the correct opening in the flaring bar, with the end of the tube extending from the tapered side of the bar the same distance as the thickness of the ring on the forming die. ☐ Task completed
11. Place the pin of the forming die into the tube and place the flaring clamp over the die and around the flaring bar. ☐ Task completed
12. Tighten the flaring clamp until the cone-shaped anvil contacts the die. Continue to tighten the clamp until the forming die contacts the flaring bar. ☐ Task completed
13. Loosen the clamp and remove the forming die. The end of the tubing should be mushroomed. ☐ Task completed

14. Place the cone-shaped anvil of the clamp into the mushroomed end of the tubing. Be careful to center the tip of the cone and verify that it is touching the inside diameter of the tubing evenly. ☐ Task completed

15. Tighten the clamp steadily until the lip formed in the first step completely contacts the inner surface of the tubing. ☐ Task completed

16. Loosen and remove the clamp and remove the tubing from the flaring bar. Inspect the flare to be sure it has the correct shape. If it is formed unevenly or cracked, you must cut off the end of the tubing and start over again. ☐ Task completed

17. To form an ISO flare, clamp the ISO flaring tool in a bench vise. Select the proper size collet and forming mandrel for the diameter of steel tubing being used. ☐ Task completed

18. Insert the mandrel into the body of the flaring tool. Hold the mandrel in place with your finger and thread in the forcing screw until it contacts and begins moving the mandrel. After contact is felt, turn the forcing screw back one full turn. ☐ Task completed

19. Slide the clamping nut over the tubing and insert the tubing into the correct collet. Leave about 3/4 in. of tubing extending out of the collet. ☐ Task completed

20. Insert the assembly into the tool body so that the end of the tubing contacts the forming mandrel. Tighten the clamping nut into the tool body very tightly to prevent the tubing from being pushed out during the forming process. Turn in the forcing screw until it bottoms out. ☐ Task completed

21. Back the clamping nut out of the flaring tool body and disassemble the clamping nut and collet assembly. Remove the tube and inspect the flare. ☐ Task completed

Instructor's Comments

____

____

____

BRAKES JOB SHEET 6

Brake Fluid Handling and Replacement

Name ______________________ Station ______________ Date ______________

NATEF Correlation

This Job Sheet addresses the following NATEF tasks:

B.8. Select, handle, store, and fill brake fluids to proper level.

B.13. Flush hydraulic system.

Objective

Upon completion of this job sheet, you will be able to select, handle, store, and install brake fluids and flush the entire brake system.

Tools and Materials

Service manual

Protective Clothing

Goggles or safety glasses with side shields

Describe the vehicle being worked on:

Year ________ Make ____________________ Model ____________________

VIN ____________________ Engine type and size ____________________

PROCEDURE

1. Follow these guidelines when working with brake fluid, especially when replacing the fluid in an entire brake system.
   a. The DOT rating is found on the container of brake fluid. The vehicle's service manual and owner's manual specify what rating is correct for the car. Always use the recommended fluid.
   b. DOT 3 and DOT 4 polyglycol fluids have a very short storage life. As soon as a container of DOT 3 or DOT 4 fluid is opened, it should be used completely because it immediately starts to absorb moisture from the air.
   c. Always store brake fluid in clean, dry containers. Brake fluid is hygroscopic; it will attract moisture and must be kept away from dampness in a tightly sealed container. Never reuse brake fluid.
   d. When working with brake fluid, do not contaminate it with petroleum-based fluids, water, or any other liquid. Keep dirt, dust, or any other solid contaminant away from the fluid.
   e. Brake fluid can cause permanent eye damage. Always wear eye protection when handling brake fluid. If you get brake fluid in your eye, see a doctor immediately.
   f. Brake fluid also will irritate your skin. If fluid gets on your skin, wash the area thoroughly with soap and water.
   g. Never mix glycol-based and silicone-based fluids because the mixture could cause a loss of brake efficiency and possible injury.

h. Do not spill glycol-based brake fluid on painted surfaces. Glycol-based fluids will damage a painted surface. Always flush any spilled fluid immediately with cold water.

i. Brake fluid is a toxic and hazardous material. Dispose of used brake fluid in accordance with local regulations and EPA guidelines. Do not pour used brake fluid down a wastewater drain or mix it with other chemicals awaiting disposal.

j. Silicone DOT 5 fluid should never be used in an antilock brake system.

2. What type of fluid is specified for this vehicle?

____

____

3. Refer to the service manual and determine whether or not the manufacturer recommends periodic brake fluid replacement. Record the recommendations and service intervals.

____

____

____

____

4. Flushing is done at each bleeder screw in the same way as bleeding the brakes. Open the bleeder screw approximately 1-1/2 turns and force fluid through the system until the fluid emerges clear and uncontaminated. ☐ Task completed

5. Repeat this at each bleeder screw in the system. ☐ Task completed

6. After all lines have been flushed, bleed the system. ☐ Task completed

Instructor's Comments

____

____

____

BRAKES JOB SHEET 7

Servicing Hydraulic System Valves

Name ______________________ Station ______________ Date ______________

NATEF Correlation

This Job Sheet addresses the following NATEF task:

B.9. Inspect, test, and/or replace metering (hold-off), proportioning (balance), pressure differential, and combination valves.

Objective

Upon completion of this job sheet, you will be able to inspect, test, and replace metering, proportioning, pressure differential, and combination valves.

Tools and Materials

Brake pressure bleeder
Clean rags
Tubing with transparent container
Two 500-psi gauges
Two 1000-psi gauges
Denatured alcohol
Hand tools
Service manual

Protective Clothing

Goggles or safety glasses with side shields

Describe the vehicle being worked on:

Year ________ Make ____________________ Model ____________________

VIN ________________ Engine type and size ____________________

PROCEDURE

Combination Valves

1. Carefully inspect the combination valve for signs of leakage around the large nut on the proportioning end. Record your findings.

____

____

2. Check around the boot for signs of leakage. If there is only a small amount of moisture inside the boot, a defective valve is not indicated. Record your findings.

____

____

3. Combination valves are nonadjustable and nonrepairable. If a valve is defective, it must be replaced. What are your recommendations for service?

____

____

____

Metering Valves

1. Carefully inspect the metering valve for signs of leakage inside the boot on the end of the valve. Record your findings and your service recommendations.

____

____

2. During the road test, did the front wheels seem to apply prematurely or lock? __________ Why could a bad metering valve be the cause of this type of problem?

____

____

3. Have a fellow student apply the brakes gradually while you watch and feel the valve stem. ☐ Task completed

4. Did the stem move? What does this indicate?

____

____

5. Metering valves can also be checked with a pressure bleeder. Charge the bleeder tank with compressed air to about 40 psi and connect it to the master cylinder. ☐ Task completed

6. Pressurize the hydraulic system with the pressure bleeder and open a front bleeder screw. Did fluid flow from the front bleeder screw? __________ If fluid did flow from the bleeder, what would be indicated?

____

____

7. Connect a T-fitting to connect one of the 500-psi gauges to the line from the master cylinder to the metering valve. ☐ Task completed

8. With another T-fitting, connect the other 500-psi gauge to the line from the metering valve to the front brakes. ☐ Task completed

9. Have a fellow student apply the brakes gradually but firmly while you watch the gauges. Record what you observed. If the valve is okay, the gauge readings should be as follows:

   a. As pressure is first applied, the readings of both gauges should increase together until the closing pressure of the valve is reached. This should be from 3 to 30 psi, depending on valve design.

b. Above the closing pressure of the valve, the inlet pressure should continue to increase while the outlet pressure stays constant.
c. As inlet pressure continues to increase, the valve will reopen. This should be from 75 to 300 psi, depending on valve design. At that point, the reading on the outlet gauge should rise to match the reading on the inlet gauge. Both gauges should then read the same as pressure continues to rise.

____

____

____

10. Metering valves are not adjustable or repairable. If a valve is defective, replace it. What are your service recommendations?

____

____

____

11. When replacing the valve, make sure to mount the new valve in the same position as the old valve. ☐ Task completed

Proportioning Valves

1. How many proportioning valves is the vehicle equipped with? And where are they located?

____

____

____

2. During the road test, did the rear wheels seem to apply prematurely or lock? __________ Why could a bad proportioning valve be the cause of this type of problem?

____

____

3. Check the valve for signs of leakage and record your findings and service recommendations.

____

____

4. How is the hydraulic system split on this vehicle?

____

____

5. If the proportioning valve has fittings on the inlet and outlet lines that allow you to connect pressure gauges with T-fittings, you can test the valve in the same way as a metering valve. If you can do this, do it, and record your results.

____

____

____

6. Refer to the service manual and find the split point of the proportioning valve. What is the split point of this proportioning valve? What does the split point tell you?

____

____

7. Connect one 1000-psi gauge to the proportioning valve inlet pressure port by using a T-fitting to connect one gauge directly to the line from the master cylinder to the proportioning valve inlet port, *or,* if you cannot connect a T-fitting to the valve inlet line, remove a bleeder screw and connect the gauge to the bleeder port of one front caliper. ☐ Task completed

8. Connect the other gauge to the proportioning valve outlet pressure port by using another T-fitting to connect the other gauge directly to the line from the proportioning valve to the rear brakes, *or,* if you cannot connect a T-fitting to the valve outlet line, remove a bleeder screw and connect the gauge to the bleeder port of one rear wheel cylinder. ☐ Task completed

   NOTE: *If a diagonally split hydraulic system has two proportioning valves, the gauge must be connected twice, and the test repeated with that hookup—once to each wheel cylinder.*

9. Have a fellow student apply the brakes gradually but firmly while you watch the gauges. Record what you observed. If the valve is okay, the gauge readings should be as follows:
   a. The readings of both gauges should increase together until the split point pressure of the valve is reached.
   b. When the pressure is above the split point, the outlet pressure should increase slower than the inlet pressure.

____

____

____

10. What are your service recommendations?

____

____

____

11. On some vehicles, proportioning valves are built into the master cylinder. These valves often can be serviced with reconditioning kits. To begin the repair, remove the master cylinder reservoir, the proportioning valve caps, and the cap O-rings. Discard the O-rings. ☐ Task completed

12. Using needle nose pliers, remove the proportioning valve piston springs and valve pistons. Be careful not to damage or scratch the piston stems. Remove the valve seals from the pistons. ☐ Task completed

13. Wash all parts with clean denatured alcohol and dry them with unlubricated compressed air. Inspect the pistons for corrosion and damage. Record your findings and service recommendations.

    ____

    ____

14. Lightly lubricate the new proportioning valve cap O-rings, valve seals, and piston stems with the special lubricant supplied in the reconditioning kit. ☐ Task completed

15. Install the valve seals on the pistons so that the seal lips are facing upward toward the caps. ☐ Task completed

16. Install the pistons and seals into the master cylinder body, followed by the valve springs. ☐ Task completed

17. Place the new cap O-rings into the grooves in the proportioning valve caps and install the valve caps, torquing them to specifications. What are the specifications? ________________

18. Reinstall the master cylinder reservoir. ☐ Task completed

Instructor's Comments

____

____

____

BRAKES JOB SHEET 8

Servicing Height-Sensing Proportional Valves

Name ______________________ Station ______________ Date ______________

NATEF Correlation

This Job Sheet addresses the following NATEF task:

B.10. Inspect, test, replace and adjust height (load) sensing proportioning valve.

Objective

Upon completion of this job sheet, you will be able to correctly inspect, test, replace, and adjust height (load) sensing proportioning valve.

Tools and Materials

Drive-on hoist

¼-inch tubing or specific gauge

Hand tools

Service manual

Protective Clothing

Goggles or safety glasses with side shields

Describe the vehicle being worked on:

Year ________ Make ____________________ Model ____________________

VIN ________________ Engine type and size ____________________

PROCEDURE

1. Describe how a height-sensing proportioning valve works.

____

____

____

2. Take a test ride. Pay attention to the action of the front and rear brakes. Do the brakes seem balanced?

____

____

3. Remove any heavy non-vehicle objects from the vehicle. What did you need to remove?

____

____

____

4. Raise the vehicle and inspect the valve for signs of leakage and damage. Record your findings and service recommendations.

____

____

5. Inspect the rear springs to see if they are damaged or have been modified. Record your finding and state why modifications can affect the operation of the proportioning valve.

____

____

____

6. If the balance between the front and rear brakes seems to be off, the height-sensing proportioning valve can be adjusted. Locate the procedure for doing this in the service manual and summarize the procedure.

____

____

____

____

7. Typically, the procedure begins with the vehicle on a lift or an alignment rack so that its wheels are on a flat surface and the vehicle is at its normal ride height. ☐ Task completed

8. Back off the valve adjuster setscrew but do not change the position of the upper nut. ☐ Task completed

9. Cut the length of 1/4-inch tubing to the adjustment specification length and slit the tubing lengthwise so you can install it on the operating rod. ☐ Task completed

10. Slip the tubing onto the operating rod. ☐ Task completed

11. With the adjuster sleeve resting on the lower mounting bracket, tighten the adjusting setscrew to lock the height set by the tube. ☐ Task completed

12. Remove the tubing from the operating rod. ☐ Task completed

13. Test drive the vehicle and check the balance of the brakes. Record your findings.

____

____

____

14. If it appears that the rear braking pressures are too little or too great, slight adjustments can be made. To do this, set the vehicle at its normal ride height. ☐ Task completed

15. Loosen the adjuster setscrew and move the adjuster sleeve toward or away from the brake pressure control valve. Move the adjuster sleeve away from the valve body on the operating rod to increase braking pressure and toward the valve body to decrease braking pressure. What did you do?

____

____

____

16. When the setting is properly adjusted, tighten the setscrew. ☐ Task completed

17. If adjustments do not correct the balance or if the valve was damaged or is leaking, it must be replaced. Begin by disconnecting the brake lines to the valve's valve body. ☐ Task completed

18. Cap the lines and tag them so that you will know where to connect them later. ☐ Task completed

19. Remove the bolt that fastens the valve bracket to the rear suspension arm and bushing. ☐ Task completed

20. Remove the bolts that secure the valve bracket to the underbody and remove the assembly. ☐ Task completed

21. Before installing the new valve, make sure the rear suspension is in full rebound. ☐ Task completed

22. If the new valve has a red plastic gauge clip, make sure it is in position on the proportioning valve and that the operating rod's lower adjustment screw is loose. ☐ Task completed

23. Position the valve and install the bolts that hold it to the underbody. ☐ Task completed

24. Install the mounting bracket to the rear suspension arm and bushing and tighten all fasteners to specifications. What are the specs? ____________

25. Make sure the valve adjuster sleeve is resting on the lower bracket and then tighten the lower adjuster setscrew. ☐ Task completed

26. Reconnect the brake lines to the correct bores in the valve body and bleed the rear brakes. ☐ Task completed

27. Remove the plastic gauge clip and lower the vehicle to the ground. ☐ Task completed

Instructor's Comments

____

____

____

BRAKES JOB SHEET 9

Brake Electrical and Electronic Component Service

Name ______________________ Station ______________ Date ______________

NATEF Correlation

This Job Sheet addresses the following NATEF tasks:

B.11. Inspect, test, and replace components of brake warning light system.

F.5. Check operation of parking brake indicator light system.

F.6. Check operation of brake stop light system; determine necessary action.

Objective

Upon completion of this job sheet, you will be able to inspect, test, and replace components of the brake stop lamp, parking brake indicator light, and brake warning light systems.

Tools and Materials

DMM

Jumper wires

Basic hand tools

Protective Clothing

Goggles or safety glasses with side shields

Describe the vehicle being worked on:

Year ________ Make ____________________ Model ____________________

VIN ____________________ Engine type and size ____________________

PROCEDURE

Stoplamps

1. Check the operation of the stoplamps. Don't forget the center mounted lamp assembly. What did you find?

____

____

2. If one of the stoplamps works but the others don't, you know you have electric power through the fuses and switches to the rear of the car. You also know the most probable cause is a bad light bulb. Sometimes, an open circuit in that part of the system may cause the same problem. Replace the bulb regardless. If the bulb was not burned out, use a DMM to locate the open circuit. Don't forget to check the ground circuit and the power circuit.

____

____

3. If all stoplamps do not work, check the circuit fuse. What did you find?

____

____

4. Locate the brake light switch on the brake pedal assembly or in the hydraulic circuit. Disconnect the harness connector and connect a jumper wire between the two terminals of the connector. ☐ Task completed

5. Check the stoplamps; they should light. If they light, replace or adjust the switch. If the lamps do not light, continue testing for an open circuit condition between the switch and the lamps. What did you find?

____

____

6. If the stoplamps work with or without depressing the brake pedal, disconnect the wiring to the switch and check the stoplamps. If the lamps are still lit, locate and repair the short circuit to battery voltage in the wiring harness. If the lamps turn off, adjust or replace the switch. What did you find?

____

____

7. If the switch needs adjustment, refer to the service manual for the correct procedure for making the adjustment before continuing this job sheet. ☐ Task completed

8. To adjust a switch that has a threaded shank and locknut, disconnect the electrical connector and loosen the locknut. Then connect an ohmmeter or a self-powered test light to the switch. Screw the switch in or out of its mounting bracket until the ohmmeter or the test light indicates continuity with the brake pedal pressed down about 1/2 inch. Tighten the locknut and reconnect the electrical harness. ☐ Task completed

9. If the switch is adjusted with a spacer or feeler gauge, loosen the switch mounting screw. Press the brake pedal and let it return freely. Then place the spacer gauge between the pedal arm and the switch plunger. Slide the switch toward the pedal arm until the plunger bottoms on the gauge. Tighten the mounting screw, remove the spacer, and check stoplamp operation. ☐ Task completed

10. If the switch has an automatic adjustment mechanism, insert the switch body in its mounting clip on the brake pedal bracket and press it in until it is fully seated. Then pull back on the brake pedal to adjust the switch position. You will hear the switch click as it ratchets back in its mount. Repeat this step until the switch no longer clicks in its mounting clip. Finally check stoplamp operation to be sure they go on and off correctly. ☐ Task completed

11. On some vehicles the stoplamp bulbs can be replaced without removing the lens cover. Simply remove the bulb and socket by twisting the socket slightly and pulling it out of the lens assembly. Push in on the bulb and rotate it. When the lugs align with the channels of the socket, pull the bulb out to remove it. ☐ Task completed

12. On other vehicles, the complete tail lamp lens assembly must be removed before replacing the bulbs. The lens assembly is typically held in position by several nuts or special screws. ☐ Task completed

Brake Warning Lamp Circuit

1. Turn the ignition switch to the start position or point halfway between on and start. The warning lamp should light. What did you find?

____

____

2. Release the ignition switch to the run position. With the parking brake off, the warning lamp should turn off. What did you find?

____

____

3. With the ignition on, apply the parking brake. The warning lamp should light. What did you find?

____

____

4. Release the parking brake. The warning lamp should turn off. What did you find?

____

____

5. If the warning lamp did what it was supposed to do in the above checks, the system is working properly. If the warning lamp stays on with the ignition on and the parking brake off, the cause of the problem is a hydraulic leak or failure in one half of the hydraulic system, low fluid level in the master cylinder reservoir, an ABS problem, or the parking brake is not fully released. ☐ Task completed
6. If the bulb never lit, there is a good chance that it is bad. Replace it. If the new bulb still doesn't work, check for power to the bulb and check the ground circuit. ☐ Task completed
7. If other indicator lamps are not operating properly, check the fuses. Then check for voltage at the last common connection. If no voltage is present here, trace the circuit back to the battery. If voltage is found at the common connection, test each branch of the circuit in the same manner. ☐ Task completed

Parking Brake Switch

1. Check the operation of the parking brake lamp, if separate from the brake warning light. The lamp should come on with the parking brake applied and go off when the brake is released. What did you find?

____

____

2. If the brake warning lamp does not light with the ignition on and the parking brake applied, the parking brake switch or circuit is faulty. To check the parking brake switch and its wiring, locate the switch on the pedal, lever, or handle and disconnect it with the ignition off. If the connector has a single wire, connect a jumper from that wire to ground. If the switch connector has two wires, connect a jumper across the two wires in the connector. Turn the ignition on and check the brake warning lamp. If the lamp is lit, replace the parking brake switch. If the lamp is still off, find and repair the open circuit in the wiring harness between the lamp and the switch. ☐ Task completed

Instructor's Comments

____

____

____

BRAKES JOB SHEET 10

Pressure Bleed Hydraulic System

Name ______________________ Station ______________ Date ______________

NATEF Correlation

This Job Sheet addresses the following NATEF task:

B.12. Bleed (manual, pressure, vacuum or surge) brake system.

Objective

Upon completion of this job sheet, you will be able to correctly bleed brake systems.

Tools and Materials

Bleeder hose
Brake fluid
Clean rag
Fender covers
Glass jar
Pressure bleeder and adapter
Pressure bleeder service manual

Protective Clothing

Goggles or safety glasses with side shields

Describe the vehicle being worked on:

Year ________ Make ____________________ Model ____________________

VIN ____________________ Engine type and size ____________________

Describe general condition:

____

PROCEDURE

CAUTION: *Be careful to avoid spills and splashes. Brake fluid can remove paint as well as do serious damage to your eyes.*

1. Place fender covers over the fenders. Study operation of the pressure bleeder in its service manual. ☐ Task completed
2. Bring the unit to a working pressure of 15–20 psi. Be sure enough brake fluid is in the pressure bleeder to complete the bleeding operation. ☐ Task completed
3. Use a rag to clean the master cylinder cover. Remove it. Remove the reservoir diaphragm gasket if there is one. Clean the gasket seat and fill the reservoir. Attach the master cylinder bleeder adapter to the reservoir following the manufacturer's instructions. ☐ Task completed
4. If the rear wheel cylinders (secondary brake system) are to be bled, use a ☐ Not applicable

suitable box wrench on the bleeder fitting at the right rear brake wheel cylinder. Attach the bleeder tube snugly around the bleeder fitting. Open the valve on the bleeder tank to admit pressurized brake fluid into the master cylinder reservoir. ☐ Task completed

5. Submerge the free end of the tube into a container partially filled with clean brake fluid, and loosen the bleeder fitting. ☐ Task completed

6. When air bubbles cease to appear in the fluid at the submerged end of the bleeder tube, close the bleeder fitting. Remove the tube. Replace the rubber dust cap on the bleeder screw. ☐ Task completed

7. Attach the bleeder tube and repeat steps 3 and 4 at the left rear wheel cylinder. ☐ Task completed

8. On the front brakes, repeat steps 3 and 4 starting at the right front disc caliper and ending at the left front disc caliper. The metering valve may require being locked in an open position. Be sure to remove the lock when the task is completed. ☐ Task completed

9. When the bleeding operation is complete, close the bleeder tank valve, and remove the tank hose from the adapter fitting. ☐ Task completed

10. After disc brake service, ensure the disc brake pistons have returned to their normal positions and the shoe and lining assemblies are properly seated. This is accomplished by depressing the brake pedal several times until normal pedal travel is established. ☐ Task completed

11. Remove the pressure bleeder adapter tool from the master cylinder. Fill the master cylinder reservoir to the "max" line or 1/4 inch below the top. Install the master cylinder cover and gasket. Ensure that the diaphragm-type gasket is properly positioned in the master cylinder cover. ☐ Task completed

CAUTION: *To prevent the air in the pressure tank from getting into the lines, do not shake the tank while air is being added to the tank or after it has been pressurized. Set the tank in the required location, bring the air hose to the tank, and do not move it during the bleeding operation. The tank should be kept at least one-third full.*

Problems Encountered

___

___

___

Instructor's Comments

___

___

___

BRAKES JOB SHEET 11

Diagnosing Drum Brake Problems

Name ______________________________ Station __________________ Date __________________

NATEF Correlation

This Job Sheet addresses the following NATEF task:

C.1. Diagnose poor stopping, noise, pulling, grabbing, dragging or pedal pulsation concerns; determine necessary action.

Objective

Upon completion of this job sheet, you will be able to diagnose poor stopping, noise, pulling, grabbing, dragging or pedal pulsation problems.

Tools and Materials

Basic hand tools

Protective Clothing

Goggles or safety glasses with side shields

Describe the vehicle being worked on:

Year __________ Make __________________________ Model __________________________

VIN __________________________ Engine type and size __________________________

PROCEDURE

1. Begin the inspection of the drum brake system by checking the tires for excessive or unusual wear or improper inflation. What did you find?

____

____

2. Wheels for bent or warped rims. What did you find?

____

____

3. Wheel bearings for looseness or wear. What did you find?

____

____

4. Suspension system for worn or broken components. What did you find?

____

____

5. Brake fluid level in the master cylinder. What did you find?

____

____

6. Signs of leakage at the master cylinder, in brake lines or hoses, at all connections, and at each wheel. What did you find?

____

____

7. Road test the vehicle. As you apply the brake pedal, check for excessive travel and sponginess. What did you find?

____

____

8. Listen for noises: not just the obvious sounds of grinding shoes or shoe linings, but mechanical clanks, clunks, and rattles. What did you find?

____

____

9. If the vehicle pulls to one side when the brakes are applied, check for glazed or worn linings, broken or weakened return spring, loose lining, inoperative self-adjuster, an out-of-round drum, or a faulty wheel cylinder at one wheel. Also check for signs of grease or brake fluid that may have contaminated the shoes and drum. Check for distorted or damaged brake pads. Grabbing brakes also may be caused by grease or brake fluid contamination. What did you find?

____

____

10. Remove the drums and inspect the brakes. Any wear in the shoes, shoe holddown and retracting hardware, drums, or wheel cylinder should be noted and corrected during a complete brake system overhaul. What did you find?

____

____

11. Check for worn shoes; glazed or worn linings; damaged or improperly adjusted brake shoes; damaged or improperly adjusted brake shoes; a loose backing plate; oil, grease, or brake fluid on the linings; faulty wheel cylinders; loose linings; out-of-round drums; or broken or weakened return springs. What did you find?

____

____

Instructor's Comments

____

____

____

BRAKES JOB SHEET 12

Inspect and Service Drum Brakes

Name ______________________________ Station __________________ Date __________________

NATEF Correlation

This Job Sheet addresses the following NATEF tasks:

C.2. Remove, clean (using proper safety procedures), inspect, and measure brake drums; determine necessary action.

C.4. Remove, clean, and inspect brake shoes, springs, pins, clips, levers, adjusters/self-adjusters, other related brake hardware, and backing support plates; lubricate and reassemble.

C.6. Pre-adjust brake shoes and parking brake before installing brake drums or drum/hub assemblies and wheel bearings.

Objective

Upon completion of this job sheet, you will be able to remove, clean, inspect, and measure brake drums. You will also be able to remove, clean, and inspect brake shoes, springs, pins, clips, levers, adjusters/self-adjusters, other related brake hardware, and backing support plates, as well as lubricate and reassemble them.

Tools and Materials

Lift or jack with jack stands
Chalk
Hand tools
Dust cap pliers
Brake adjusting tool
Vacuum brake cleaner
Brake cleaning equipment
Vernier caliper or depth micrometer
Brake drum micrometer
Return spring pliers
Hold-down spring tool
Brake shoe adjusting gauge
Service manual

Protective Clothing

Goggles or safety glasses with side shields

Describe the vehicle being worked on:

Year __________ Make ________________________ Model ________________________

VIN ____________________ Engine type and size ________________________

PROCEDURE

1. Does the vehicle have ABS? ____________ If yes, summarize the precautions that must be followed to safely work on the brakes. Then do them.

   ____

   ____

   ____

2. Check the fluid level in the master cylinder and adjust the level so the reservoir is about half full. ☐ Task completed
3. Disconnect the battery negative cable. If the vehicle has electronically controlled suspension, turn the suspension service switch off. ☐ Task completed
4. Raise the vehicle on a hoist or safety stands and remove the wheels. ☐ Task completed
5. Remove the wheels from the brakes to be serviced. Remember, brakes are always serviced in axle sets. ☐ Task completed
6. Vacuum or wet-clean the brake assembly to remove all dirt, dust, and fibers. If you are wet-cleaning the brakes, position a catch basin under the brake units. ☐ Task completed
7. Inspect the brake drum mounting and identify any fasteners that may need to be removed before the drum can be pulled off. List them here.

   ____

   ____

8. Lightly pull on the drum. Does it feel like it will come off easily? ________
9. If the drum seems tight, you may have to back off the parking brake adjustment or manually retract the self-adjusters to gain shoe-to-drum clearance for drum removal. What did you do?

   ____

   ____

10. Mark the outside of the drum with a "L" or "R" to show what side the drum came from. Then, remove the drum. ☐ Task completed
11. Inspect the drum for damage and defects. Record your findings.

    ____

    ____

12. Locate the specifications for brake drums in the service manual and record them here.

    ____

    ____

13. Measure the inside diameter of the drum and compare them to the specifications.

    ____

    ____

14. What are your service recommendations for the drums?

____

____

____

15. Thoroughly clean the brake assembly, including the backing plates, struts, levers, and other metal parts that will be reused; make sure all liquid lands in the catch basin. ☐ Task completed

16. Examine the rear wheels for evidence of oil or grease leakage past the wheel bearing seals. ☐ Task completed

17. Inspect the shoes and lining, paying attention to the lining's thickness, wear pattern, and damage. Also inspect the lining for cracks, missing rivets, and looseness. Check for contamination from grease, oil, or brake fluid. Record your findings.

____

____

18. Check the linings for unequal wear on any shoe of an axle set. Also look for uneven lining wear on any particular shoe. Record your findings.

____

____

19. Inspect the outside of the wheel cylinder for leakage. Then pull back the dust boots and look for fluid at the ends of the cylinder. Record your findings.

____

____

20. Check the cylinder mounting on the backing plate for looseness and missing fasteners. Record your findings.

____

____

21. Check the backing plate for breaks, distortion, or other damage. Also look for uneven wear on the shoe support pads that could indicate a bent shoe or incorrectly installed parts. Record your findings.

____

____

22. Inspect the return and hold-down springs for damage and unusual wear. Pry the brake shoes slightly away from the backing plate and release them. The hold-downs should pull the shoes sharply back to the plate. Record your findings.

____

____

23. Inspect the self-adjuster levers and pawls for wear and damage. Record your findings.

____

____

24. Check the parking brake linkage for damage and rust. Be sure that all parking brake levers and links are free to move easily. Record your findings.

____

____

25. Select the correct return spring tool and remove the top return spring from its anchor. Note how they are hooked over anchor posts and to which holes in the shoe webs they are attached. ☐ Task completed

26. Remove the hold-down springs and clips. ☐ Task completed

27. Remove shoe-to-shoe springs and self-adjuster assemblies. Make sure you note the positions in which the parts are installed so that you can reinstall them correctly. ☐ Task completed

28. Disconnect the parking brake linkage from the brake shoes. ☐ Task completed

29. Remove the brake shoes. ☐ Task completed

30. Compare the new brake shoes to the old ones to make sure the holes for springs and clips are in the same locations and that the new linings are positioned in the same locations on the shoes. ☐ Task completed

31. Gather all new parts for the assembly and check to make sure they are good replacements. ☐ Task completed

32. Transfer any parking brake linkage parts from the old shoes to the new shoes. ☐ Task completed

33. Remove all nicks and rough spots from the shoe pads on the backing plate with emery cloth and then clean the area. ☐ Task completed

34. Lightly coat the shoe pads with brake lubricant. ☐ Task completed

35. Mount the shoes and install the hold-down parts. ☐ Task completed

36. Lightly coat the surface of the parking brake pin with brake lubricant. Install the lever on the pin with a new washer and clip. Attach the parking brake cables and make sure their movement is not restricted. ☐ Task completed

37. Make sure the shoe webs fit into the ends of the wheel cylinder pistons correctly. ☐ Task completed

38. Disassemble the brake star adjuster and clean the parts in denatured alcohol. ☐ Task completed

39. Clean the threads of the adjuster with a fine wire brush and make sure the entire length of the adjuster threads into the threaded sleeve without sticking or binding. ☐ Task completed

40. Check the starwheel teeth for damage. Then, lubricate the threads with brake lubricant, being careful not to get any on the starwheel teeth. ☐ Task completed

41. Set the star adjusters to their most retracted position, and then install them. ☐ Task completed

42. After the self-adjusters are installed, check their operation by lightly prying the shoe away from its anchor or by pulling the cable or link to make sure the adjuster easily advances one notch at a time. ☐ Task completed

43. Use brake spring pliers and other special tools to install the return springs. Make sure each spring is heading in the proper direction and is in the proper holes in the shoe webs. ☐ Task completed

44. Check the parking brake cable adjustment. The parking brake must operate freely with the brake shoes and linings centered on the backing plate. ☐ Task completed

45. Place the brake shoe adjusting gauge into the drum and slide it back and forth until the jaws are open to their widest point. Then tighten the lockscrew to hold the jaws in position. ☐ Task completed

46. Turn over the adjusting gauge and place that side over the shoes. ☐ Task completed

47. Rotate the starwheel adjuster to expand the shoes until the caliper just slides over them without binding. ☐ Task completed

48. Install the brake drum and apply the brakes several times to verify that the pedal is fairly high and firm. Bleed the brakes, if necessary, to ensure that the lines are free from air. Then make a final, manual adjustment of the brakes. ☐ Task completed

49. Install the wheels and torque the lug nuts to specifications. What are the specs? ________________

50. Reconnect the battery negative cable. If the vehicle has electronically controlled suspension, turn the suspension service switch back on. ☐ Task completed

51. Check the fluid level in the master cylinder and add fluid to bring it to the proper level. ☐ Task completed

Instructor's Comments

____

____

____

BRAKES JOB SHEET 13

Refinishing Brake Drums

Name ______________________ Station ______________ Date ______________

NATEF Correlation

This Job Sheet addresses the following NATEF task:

C.3. Refinish brake drum.

Objective

Upon completion of this job sheet, you will be able to mount a brake drum on a brake lathe and machine the braking surface.

Tools and Materials

Brake lathe Emery cloth
Shop rags #80 Sandpaper

Protective Clothing

Goggles or safety glasses with side shields

Describe the vehicle being worked on:

Year ________ Make ____________________ Model ____________________

VIN ____________________ Engine type and size ____________________

PROCEDURE

NOTE: *Brake drums should be refinished only under the following conditions:*

- If the inside diameter of the drum is great enough to allow metal removal and still be above the minimum wear dimension.
- If it is bell-mouthed, concave or convex, or out-of-round.
- If there is noticeable brake pulsation.
- If there are heat spots or excessive scoring that can be removed by resurfacing.

1. Here are guidelines to adhere to when refinishing drums:
   - Whenever you refinish a drum, remove the least amount of metal to achieve the proper finish.
   - Never turn the drum on one side of the vehicle without turning the drum on the other side. Drum diameters must be within 0.005 inch of one another.
   - Clean any oil film off the drum with brake cleaning solvent or alcohol and let the drum air dry before installing it on the vehicle or the lathe.
   - Some replacement brake drums are semi-finished and may require additional machining to obtain the proper dimensional specifications and surface finish. Fully finished drums do not require additional machining unless it is needed to match the diameter of an old drum on the same axle.

2. Describe the type of lathe you will be using for this job sheet.

____

____

3. Check the lathe for condition, as well as the attaching adapters, tool holders, vibration dampers, and cutting bits. Make sure the mounting adapters are clean and free of nicks. Always use sharp cutting tools or bits and use only replacement cutting bits recommended by the equipment manufacturer. The tip of the cutting bit should be slightly rounded. State the results of this check.

____

____

4. Mount the drum onto the lathe. A one-piece drum with bearing races in its hub mounts to the lathe arbor with tapered or spherical cones. A two-piece drum removed from its hub is centered on the lathe arbor with a spring-loaded cone and clamped in place by two large cup-shaped adapters. ☐ Task completed

5. Remove all grease and dirt from the bearing races before mounting the drum. Make sure the bearing races are not loose in the drum. Use the appropriate cones and spacers to lock the drum firmly to the arbor shaft. When mounting a two-piece drum without the hub, clean all rust and corrosion from the hub area with emery cloth. Use the proper cones and spacers to mount the drum to the arbor shaft. ☐ Task completed

6. After the drum is on the lathe, install the vibration damper band on the outer diameter of the drum to prevent the cutting bits from chattering during refinishing. ☐ Task completed

7. Before removing any metal from the drum, verify that it is centered on the lathe arbor. ☐ Task completed

8. Make a small scratch on one surface of the drum to check drum mounting. Do this by backing the cutting assembly away from the drum and turning the drum through one complete revolution to be sure there is no interference with rotation. ☐ Task completed

9. Start the lathe and advance the cutting bit until it just touches the drum surface near midpoint. ☐ Task completed

10. Let the cutting bit lightly scratch the drum, approximately 0.001 inch (0.025 mm) deep. ☐ Task completed

11. Move the cutting bit away from the drum and stop the lathe. If the scratch is all the way around the drum, the drum is centered and you can proceed with resurfacing. ☐ Task completed

12. If the scratch appears intermittently, either the drum is out of round or it is not centered on the arbor. In this case, loosen the arbor nut and rotate the drum 180 degrees on the arbor; then retighten the nut. ☐ Task completed

13. Repeat steps 9 through 11 to make another scratch about 1/4 inch away from the first. ☐ Task completed

14. If the second scratch appears intermittently, the drum is significantly out of round, but it is properly centered on the lathe and you can proceed with machining. ☐ Task completed

15. If the second scratch appears opposite the first on the drum surface, remove the drum from the lathe arbor and recheck the mounting. ☐ Task completed

16. Turn on the lathe and adjust it to the desired speed. Then advance the cutting bit to the open edge of the drum and remove the ridge of rust and metal that has formed there. Use several light cuts of 0.010 to 0.020 inch rather than one heavy cut. ☐ Task completed

17. Move the cutting bit to the closed (inner) edge of the drum and remove the ridge that also is present there. As you remove the ridges, note the point of the smallest drum diameter. Position the cutting bit at this point and adjust the handwheel that controls the depth of cut to zero. This is the starting point for further depth-of-cut adjustments. ☐ Task completed

18. Reposition the cutting bit to the closed (inner) edge of the drum and adjust it for a rough cut. The handwheel micrometer is graduated to indicate the amount of metal removed from the complete diameter. The actual depth of the cut is one-half of that setting. ☐ Task completed

19. Adjust the crossfeed for a rough cut and engage the crossfeed mechanism. The lathe will automatically move the cutting bit from the inner to the outer edge of the drum. Make as many rough cuts as necessary to remove defects, but stay within the dimension limits of the drum. ☐ Task completed

20. Complete the turning operation with a finish cut. ☐ Task completed

21. After refinishing, keep the drum mounted on the lathe, and deburr it with no. 80 grit sandpaper to remove all minor and small rough and jagged surfaces. ☐ Task completed

22. After the drum has been sanded, clean each surface with hot water and detergent. Then dry the drum thoroughly with clean, lint-free paper towels and allow it to air dry. ☐ Task completed

Instructor's Comments

____

____

____

BRAKES JOB SHEET 14

Wheel Cylinder Service

Name ______________________ Station ______________ Date ______________

NATEF Correlation

This Job Sheet addresses the following NATEF task:

C.5. Remove, inspect, and install wheel cylinders.

Objective

Upon completion of this job sheet, you will be able to remove, inspect, and install wheel cylinders.

Tools and Materials

Basic hand tools	Shop towels
Line wrenches	Brake cleaning solvent

Protective Clothing

Goggles or safety glasses with side shields

Describe the vehicle being worked on:

Year ________ Make ____________________ Model ____________________

VIN ____________________ Engine type and size ____________________

PROCEDURE

1. To service a wheel cylinder, the brake drum must be removed. ☐ Task completed
2. Check for leaks at or around the wheel cylinder; look for fluid on the cylinder, brake linings, and backing plate, and on the inside of the tire. What did you find?

   ____

   ____

3. Gently peel back the cylinder's dust boot and look for signs of fluid. A small amount of fluid seepage dampening the interior of the boot is normal. A dripping boot is not. What did you find?

   ____

   ____

4. Since brake hoses are an important link in the hydraulic system, it is recommended that they be replaced when a new cylinder is to be installed. Remove the brake shoe assemblies from the backing plate before proceeding. Make sure brake fluid or grease does not get on the brake lining. ☐ Task completed

5. Using two appropriate wrenches, disconnect the hydraulic hose from the steel line located on the chassis. On solid rear axles, use the appropriate tubing wrench and disconnect the hydraulic line where it enters the wheel cylinder. Care must be exercised in removing this steel line. It might be bent at this point and be difficult to install once new wheel cylinders are mounted to the backing plate. ☐ Task completed

6. Remove the plates, shims, and bolts that hold the wheel cylinder to the backing plate. Some wheel cylinders are held to the backing plate with a retaining ring that can be removed with two small picks. ☐ Task completed

7. Remove the wheel cylinder from the backing plate and clean the area with a proper cleaning solvent. ☐ Task completed

8. Care must be taken when installing wheel cylinders on cars equipped with wheel cylinder piston stops. The rubber dust boots and the pistons must be squeezed into the cylinder before it is tightened to the backing plate. If this is not done, the pistons jam against the stops causing hydraulic fluid leaks and erratic brake performance. ☐ Task completed

9. Position the wheel cylinder with its plates, shims, and bolts, and tighten it in place. ☐ Task completed

10. Using two appropriate wrenches, connect the hydraulic hose from the steel line located on the chassis. On solid rear axles, use the appropriate tubing wrench and connect the hydraulic line where it enters the wheel cylinder. ☐ Task completed

11. Install the brake shoe assemblies onto the backing plate. Make sure brake fluid or grease does not get on the brake lining. ☐ Task completed

12. Set the width of the shoes and install the brake drum. ☐ Task completed

13. Bleed the brake system. ☐ Task completed

Instructor's Comments

____

____

____

BRAKES JOB SHEET 15

Removing and Installing a Tire and Wheel Assembly

Name ______________________ Station ______________ Date ______________

NATEF Correlation

This Job Sheet addresses the following NATEF task:

C.7. Install wheel, torque lug nuts, and make final checks and adjustments.

Objective

Upon completion of this job sheet, you will be able to install wheels, torque lug nuts, as well as make final checks and adjustments.

Tools and Materials

Breaker bar	Wheel chocks
Hydraulic floor jack	Safety stands

Protective Clothing

Goggles or safety glasses with side shields

Describe the vehicle being worked on:

Year ________ Make ____________________ Model ____________________

VIN __________________ Tire size ______________ Wheel Diameter ______

PROCEDURE

1. Determine the proper lift points for the vehicle. Describe them here:

   ____

   ____

2. Place the pad of the hydraulic floor jack under one of the lift points for the left side of the front of the vehicle. Describe this location.

   ____

   ____

3. Raise the pad just enough to make contact with the vehicle. ☐ Task completed

4. Make sure the vehicle's transmission is placed in park or in a low gear (if there is a manual transmission) and apply the parking brake. ☐ Task completed

5. Place blocks or wheel chocks at the rear tires to prevent the vehicle from rolling. ☐ Task completed

6. Remove any hub or bolt caps to expose the lug nuts or bolts. ☐ Task completed

7. Refer to the service or owner's manual to determine the order the lugs ought to be loosened. If this is not available, make sure they are alternately loosened. Describe the order you will use to loosen and remove the lug nuts.

____

____

8. Get the correct size socket or wrench for the lug nuts. The size is: ________

9. Using a long breaker bar, loosen each of the lug nuts but do not remove them. ☐ Task completed

10. Raise the vehicle with the jack. Make sure it is high enough to remove the wheel assembly. Once it is in position, place a safety stand under the vehicle. Describe the location of the stand.

____

____

11. Lower the vehicle onto the stand and lower and remove the jack from the vehicle. ☐ Task completed

12. Remove the lug nuts and the tire and wheel assembly. ☐ Task completed

13. Before installing the wheel and tire assembly, make sure the tire is inflated to the recommended inflation. What is that inflation?

____

14. Move the assembly over to the vehicle. ☐ Task completed

15. Place the assembly onto the hub and start each of the lug nuts. ☐ Task completed

16. Hand tighten each lug nut so that the wheel is fully seated against the hub. ☐ Task completed

17. Put the hydraulic jack into position and raise the vehicle enough to remove the safety stands. ☐ Task completed

18. Lower the jack so the tire is seated on the shop floor. ☐ Task completed

19. Following the reverse pattern as used to loosen the lug nuts, tighten the lug nuts according to specifications. The order you used to tighten was:

____

____

The recommended torque spec is: ______________________

20. What happens if you over tighten the lug nuts?

____

____

____

21. Install the bolt or hub cover. ☐ Task completed

Instructor's Comments

BRAKES JOB SHEET 16

Diagnosing Disc Brake Problems

Name ______________________________ Station __________________ Date __________________

NATEF Correlation

This Job Sheet addresses the following NATEF task:

D.1. Diagnose poor stopping, noise, pulling, grabbing, dragging or pedal pulsation concerns; determine necessary action.

Objective

Upon completion of this job sheet, you will be able to diagnose poor stopping, noise, pulling, grabbing, dragging or pedal pulsation problems.

Tools and Materials

Basic hand tools

Protective Clothing

Goggles or safety glasses with side shields

Describe the vehicle being worked on:

Year ___________ Make __________________________ Model __________________________

VIN __________________________ Engine type and size __________________________

PROCEDURE

1. Begin the inspection of the disc brake system by checking the tires for excessive or unusual wear or improper inflation. What did you find?

____

____

2. Wheels for bent or warped rims. What did you find?

____

____

3. Wheel bearings for looseness or wear. What did you find?

____

____

4. Suspension system for worn or broken components. What did you find?

____

____

5. Brake fluid level in the master cylinder. What did you find?

____

____

6. Signs of leakage at the master cylinder, in brake lines or hoses, at all connections, and at each wheel. What did you find?

____

____

7. Road test the vehicle. As you apply the brake pedal, check for excessive travel and sponginess. What did you find?

____

____

8. Listen for noises: not just the obvious sounds of grinding pads or pad linings, but mechanical clanks, clunks and rattles. What did you find?

____

____

9. If the vehicle pulls to one side when the brakes are applied, check for a bad caliper or loose caliper at one wheel. Also check for signs of grease or brake fluid that may have contaminated the pads and rotor. Check for distorted or damaged brake pads. Grabbing brakes also may be caused by grease or brake fluid contamination or by a malfunctioning or loose caliper. What did you find?

____

____

10. Remove the wheels and make a careful inspection of the brake pads and caliper mounting hardware. If the brakes are equipped with caliper slides, check them as well. What did you find?

____

____

11. Check for worn rotors or pads. These may also cause roughness or pedal pulsation when the brakes are applied. What did you find?

____

____

Instructor's Comments

____

____

____

BRAKES JOB SHEET 17

Inspect and Service Typical Disc Brakes

Name ______________________ Station ______________ Date ______________

NATEF Correlation

This Job Sheet addresses the following NATEF tasks:

D.2. Remove caliper assembly from mountings; clean and inspect for leaks and damage to caliper housing; determine necessary action.

D.3. Clean and inspect caliper mounting and slides for wear and damage; determine necessary action.

D.4. Remove, clean, and inspect pads and retaining hardware; determine necessary action.

Objective

Upon completion of this job sheet, you will be able to remove the caliper assembly from mountings; clean and inspect the caliper housing for leaks and damage. You will also be able to clean and inspect the caliper mounting and slides for wear and damage and remove, clean, and inspect the pads and retaining hardware.

Tools and Materials

Appropriate screwdrivers
Appropriate wrenches
Asbestos removal system
Boot ring compressor
Boot seal installer
Brake fluid
C-clamps
O-rings
Plastic sealant
Service manual
Torque wrench
Wire brush

Protective Clothing

Goggles or safety glasses with side shields
Respirator

Describe the vehicle being worked on:

Year ________ Make ____________________ Model ____________________

VIN ____________________ Engine type and size ____________________

Describe general condition:

____

PROCEDURE (CALIPER REMOVAL)

1. Raise the vehicle and remove the wheel and tire assembly. Use care to avoid damage to or interference with the bleeder screw during removal. ☐ Task completed

2. Install an asbestos removal vacuum system according to manufacturer's instructions. Remove all dust from the brake components. ☐ Task completed

3. Mark the right-hand and left-hand caliper assemblies with chalk prior to removal from the vehicle, so they can be positioned correctly during installation. ☐ Task completed

4. On a sliding or floating caliper, install a C-clamp on the caliper with the solid end of the clamp on the caliper housing and the screw end on the metal portion of the outboard brake pad. Tighten the clamp until the piston bottoms in the caliper bore. Then remove the clamp. Bottoming the piston allows room for the brake pad to slide over the ridge of rust that accumulates on the edge of the rotor. ☐ Task completed

NOTE: *First, remove 1/3 of the fluid from the master cylinder to prevent spillage and damage to painted parts and certain types of plastics.*

5. Disconnect the brake hose from the caliper and remove the copper gasket or washer and cap the end of the brake line. If only brake pads are to be replaced, do not disconnect the brake hose. ☐ Task completed

6. Remove the two mounting brackets to the steering knuckle bolts. Support the caliper when removing the second bolt to prevent the caliper from falling. ☐ Task completed

7. On a sliding caliper, remove the top bolts, retainer clip, and antirattle springs. On a floating caliper, remove the two special pins that hold the caliper to the anchor plate. On older type fixed calipers, remove the bolts holding them to the steering knuckle. On all three types, get the caliper off by prying it straight up and lifting it clear of the rotor. ☐ Task completed

PROCEDURE (BRAKE AND REMOVAL)

1. If the friction pads appear worn and in need of replacement, measure them at the thinnest part of the pad. If they are 2/32 inch thick or less, replace the pads. Refer to the disc brake torque and specification chart in the service manual for minimum brake pad lining thickness. ☐ Task completed ☐ Not applicable

2. Uneven pad wear on a sliding caliper often means that the caliper is sticking and not giving equal pressure to both pads. On a sliding caliper, the problem could be that the caliper "ways" are not allowing for a smooth sliding movement. Check these machined ways for proper clearance. A slightly tapered wear pattern on the pads of certain models is caused by caliper twist during braking. It is normal if it does not exceed 1/8 inch taper from one end of the pad to the other. ☐ Task completed ☐ Not applicable

3. Sliding or floating calipers must always be lifted off the rotor for pad replacement. Fixed calipers might have pads that can be replaced by removing the retaining pins or clips instead of having to lift off the entire caliper. They can be held in position by retaining pins, guide pins, or a support key. Note the position of the shims, antirattle clips, keys, bushings, or pins during disassembly. ☐ Task completed ☐ Not applicable

4. If only pads need to be replaced, lift the caliper off the rotor and hang it up by a wire. Remove the outer pad and inner pad. Remove the old sleeves and bushings and install new ones. Replace the rusty pins on a floating caliper to provide for free movement. Transfer the shoe retainers, which can be slips or springs, onto the new pads.
   - ☐ Task completed
   - ☐ Not applicable

PROCEDURE (CALIPER DISASSEMBLY)

1. Position the caliper face down on a bench. Insert the used outer pad into the caliper. Place a folded shop towel on the face of the lining to cushion the piston.
   - ☐ Task completed
2. Apply low air pressure (never more than 30 psi) to the fluid inlet port of the caliper to force the piston from the caliper housing.
   - ☐ Task completed

   WARNING: *Be careful to apply the air pressure very gradually, and be sure there are enough cloths to catch the piston when it comes out of the bore. Never place your fingers in front of the piston for any reason when applying compressed air. Personal injury could occur if the piston is "popped" out of the bore.*
3. If the piston is frozen, release the air pressure and tap the end of the piston with a soft-headed hammer or mallet. Reapply air pressure.
   - ☐ Task completed
   - ☐ Not applicable
4. Frozen phenolic (plastic) pistons can be broken into pieces with a chisel and hammer.
   - ☐ Task completed
   - ☐ Not applicable

   WARNING: *Always wear safety goggles or glasses with side shields when completing this task. Avoid damaging the caliper bore and seal groove with the chisel.*
5. Internal expanding pliers are sometimes used to remove the pistons from the caliper bores.
   - ☐ Task completed
   - ☐ Not applicable

PROCEDURE (CLEANING AND INSPECTING)

1. Inspect the phenolic pistons for cracks, chips, or gouges. Replace the piston if any of these conditions is evident. If the plated surface of a steel piston is worn, pitted, scored, or corroded, it should also be replaced.
   - ☐ Task completed
2. Dust boots vary in design depending on the type of piston and seal, but they all fit into one groove in the piston and the other groove in the cylinder. One type comes out with the piston and peels off. Another type stays in place and the piston comes out through the boot, and then is removed from the cylinder. In either case, peel the boot from its groove. In some cases it might be necessary to pry out the dust boot with a screwdriver. When prying a dust boot, be careful not to scratch the cylinder bore. The old boot can be discarded because it must be replaced along with the seal.
   - ☐ Task completed
   - ☐ Not applicable
3. Pry the piston seal out of the cylinder with a wooden or plastic tool. Do not
   - ☐ Task completed

use a screwdriver or other metal tool. Any of these could nick the metal in the caliper bore and cause a leak. ☐ Not applicable

4. Inspect the bore for pitting or scoring. A bore that shows light scratches or corrosion can usually be cleaned with crocus cloth. However, a bore that has deep scratches or scoring can be honed, provided the diameter of the bore is not increased more than 0.002 inch. ☐ Task completed
5. If the bore does not clean up within specifications, the new caliper housing should be installed (black stains on the bore walls are caused by piston seals and will do no harm). ☐ Task completed ☐ Not applicable
6. When using a hone, be sure to install a hone baffle before honing the bore. The baffle protects the hone stones from damage. Use extreme care in cleaning the caliper after honing. ☐ Task completed ☐ Not applicable
7. Remove all dust and grit by flushing the caliper with alcohol. Wipe it dry with a clean, lint-free cloth and then clean the caliper a second time in the same manner. ☐ Task completed

PROCEDURE (CALIPER ASSEMBLY)

1. Lubricate the new piston seal with clean brake fluid or assembly lubricant (usually supplied with the caliper rebuild kit). Make sure the seal is not distorted. Insert it into the groove in the cylinder bore so it does not become twisted or rolled. ☐ Task completed
2. Install a new dust boot by setting the flange squarely in the outer groove of the caliper bore. ☐ Task completed
3. Coat the piston with brake fluid or assembly lubricant and install it in the cylinder bore. Be sure to use a wood block or other flat stock when installing the piston back into the bore. Never apply a C-clamp directly to a phenolic piston, and be sure the pistons are not cocked. Spread the dust boot over the piston as it is installed. Seat the dust boot in the piston groove. ☐ Task completed
4. Install the bleeder screw. ☐ Task completed

PROCEDURE (BRAKE PAD INSTALLATION—FIXED CALIPER PADS)

1. Insert the new pads and plates in the caliper with the metal plates against the end of the pistons. Make sure the plates are properly seated in the caliper. ☐ Task completed

   NOTE: *Make sure the noise suppressor compound has been applied to the back of the brake pads, following the procedure outlined in the service manual.*

2. Spread the pads and slide the caliper into position on the rotor. With some pads, the mounting bolts are used to hold them in place. These bolts are usually tightened 80–90 ft-lb. On some fixed-disc brakes, the pads are held in place by retaining clips or, as with some Delco-Moraine designs, both ☐ Task completed ☐ Not applicable

retaining pins and clips. Channel lock pliers are used to clinch the ear clips of the outer pads.

3. Reinstall the antirattle spring/clips and other hardware (if so equipped). ☐ Task completed ☐ Not applicable

PROCEDURE (BRAKE PAD INSTALLATION—SLIDING CALIPER PADS)

1. Push the piston carefully back into its bore until it bottoms. Slide a new outer pad and lining assembly into the recess of the caliper. No free play between the brake pad flanges and caliper flanges should exist. If free play is found, remove the pad from the caliper and bend the flanges to eliminate all vertical free play. Install the pad. ☐ Task completed
2. Place the inner pad into position on the adapter with the pad flange in the adapter's machined ways. Slide the adapter assembly into position in the adapter and over the disc. Align the caliper on the adapter's ways. Do not pull the dust boot from its groove when the piston and boot are slid over the inboard pad. ☐ Task completed
3. Install the antirattle springs (if so equipped) on the top of the retainer plate and tighten the retaining screws to specifications. ☐ Task completed ☐ Not applicable

PROCEDURE (BRAKE PAD INSTALLATION—FLOATING CALIPER PADS)

1. Compress the flanges of the outer bushing in the caliper fingers and work them into position in the hole from the outer side of the caliper. Compress the flanges of the inner guide pin bushings and install them. ☐ Task completed
2. Slide the new pad and lining assemblies into position on the adapter and caliper. Make sure the metal portion of the pad is fully recessed in the caliper and adapter and that the proper pad is on the outer side of the caliper. ☐ Task completed
3. Hold the outer pad and carefully slide the caliper into position on the adapter and over the disc. Align the guide pin holes of the adapter with those of the inner and outer pads. ☐ Task completed
4. Install the guide pins through the bushings, caliper, adapter, and inner and outer pads into the outer bushings in the caliper and antirattle spring. ☐ Task completed

PROCEDURE (CALIPER INSTALLATION)

1. Install the caliper assembly over the rotor with the outer brake pad against the rotor's braking surface. This prevents pinching the piston boot between the inner brake pad and the piston. Make sure the correct caliper is installed on the correct anchor plate according to the way they were marked during disassembly. ☐ Task completed

2. Lubricate the rubber insulators (if so equipped) with a silicone dielectric compound. ☐ Task completed
3. Install the caliper assembly back on its mounting brackets. ☐ Task completed
4. Connect the brake hose to the caliper. If copper washers or gaskets are used, be sure to use new ones—the old ones might have taken a set and might not form a tight seal if reused. ☐ Task completed
5. Fill the master cylinder reservoirs and bleed the hydraulic system. ☐ Task completed
6. Check for fluid leaks under maximum pedal pressure. ☐ Task completed
7. Lower the vehicle and road test it. ☐ Task completed

Problems Encountered

___

___

___

Instructor's Comments

___

___

___

BRAKES JOB SHEET 18

Service a Single-Piston Disc Caliper

Name ______________________________ Station __________________ Date __________________

NATEF Correlation

This Job Sheet addresses the following NATEF tasks:

D.5. Disassemble and clean caliper assembly; inspect parts for wear, rust, scoring, and damage; replace seal, boot, and damaged or worn parts.

D.6. Reassemble, lubricate, and reinstall caliper, pads, and related hardware; seat pads, and inspect for leaks.

Objective

Upon completion of this job sheet, you will be able to disassemble and clean the caliper assembly; inspect parts for wear, rust, scoring, and damage; replace the seal, boot, and damaged or worn parts. You will also be able to reassemble, lubricate, and reinstall caliper, pads, and related hardware; seat pads, and inspect for leaks.

Tools and Materials

Appropriate screwdrivers
Appropriate wrenches
Brake fluid
C-clamps
Service manual
Torque wrench
Wire
Wood block

Protective Clothing

Goggles or safety glasses with side shields
Respirator

Describe the vehicle being worked on:

Year __________ Make _________________________ Model _________________________

VIN _________________________ Engine type and size _________________________

Describe general condition:

____

PROCEDURE

1. Remove the caliper retaining clips and antirattle springs. Rotate and lift the caliper from the anchor plate. Hang it by wire. ☐ Task completed

2. Remove the inner pad from the anchor plate. ☐ Task completed

3. Remove the rotor. (If the drum shoes prevent removal, insert an adjusting spoon through the access hole in the splash shield and back off the star wheel adjuster to release the shoes. Then pull off the rotor.) ☐ Task completed

4. Remove the lower shoe-to-shoe spring. Spread the shoes apart and remove the star wheel adjuster. Remove the upper shoe-to-shoe spring. ☐ Task completed

5. Move the shoes out of the support platform, and mark them for identification. ☐ Task completed

6. Remove the hold-down springs and shoes. ☐ Task completed

7. Inspect the brake lining; replace as necessary. ☐ Task completed

8. Reposition the brake shoes and install the hold-down springs. Attach the upper shoe-to-shoe return spring. Install the star wheel adjuster mechanism. On the left side, the star wheel is forward. On the right side, it is rearward. Attach the lower shoe-to-shoe spring. ☐ Task completed

9. Replace the rotor. ☐ Task completed

10. Seat the parking brake shoes against the drum by rotating the star wheel adjuster. Use an adjusting spoon through the splash shield access hole. Back off the star wheel 12 clicks to provide enough running clearance. ☐ Task completed

11. Install the inboard pad onto the adapter ways of the anchor plate. ☐ Task completed

12. Slide the caliper assembly slowly over the rotor until it contacts the machined abutments on the anchor plate. ☐ Task completed

13. Install the retaining clips and torque the screws to 15 ft-lb. ☐ Task completed

14. Install the antirattle springs on the top. ☐ Task completed

15. Bleed the brakes. ☐ Task completed

16. Test the pedal and road test the car. ☐ Task completed

Problems Encountered

____

____

____

Instructor's Comments

____

____

____

BRAKES JOB SHEET 19

Inspect and Measure Brake Rotors

Name ______________________ Station ______________ Date ______________

NATEF Correlation

This Job Sheet addresses the following NATEF task:

D.7. Clean, inspect, and measure rotor with a dial indicator and a micrometer; follow manufacturer's recommendations in determining need to machine or replace.

Objective

Upon completion of this job sheet, you will be able to clean, inspect, and correctly measure a rotor with a dial indicator and a micrometer.

Tools and Materials

Hand tools

Outside micrometer

Service manual

Protective Clothing

Goggles or safety glasses with side shields

Describe the vehicle being worked on:

Year ________ Make ____________________ Model ____________________

VIN ____________________ Engine type and size ____________________

PROCEDURE

1. Remove the rotor. ☐ Task completed
2. Visually inspect the rotor for damage or excessive wear, such as scoring or heat cracking. ☐ Task completed
3. Each rotor is usually stamped with a minimum thickness specification. If you cannot find this on the rotor, refer to the service manual for minimum thickness specifications. ☐ Task completed
4. Using an outside micrometer, measure the thickness of the rotor on the braking surface in three or four locations on the rotor. ☐ Task completed
5. If the rotor is not below minimum thickness and shows signs of wear, it can be resurfaced. After this procedure, the rotor thickness should be remeasured. ☐ Task completed
6. If the rotor is below minimum thickness, it must be replaced. A rotor that is below minimum thickness can be unsafe. ☐ Task completed

7. Reinstall the rotor, caliper assembly, and wheel and tire assembly. Road test the vehicle. ☐ Task completed

Problems Encountered

____

____

____

Instructor's Comments

____

____

____

BRAKES JOB SHEET 20

Refinishing Brake Rotors

Name ______________________ Station ______________ Date ______________

NATEF Correlation

This Job Sheet addresses the following NATEF task:

D.9. Refinish rotor according to manufacturer's recommendations.

Objective

Upon completion of this job sheet, you will be able to refinish brake rotors according to manufacturer's recommendations.

Tools and Materials

Brake lathe
Shop rags
Emery cloth
Sanding disc power tool
Micrometer

Protective Clothing

Goggles or safety glasses with side shields

Describe the vehicle being worked on:

Year ________ Make ____________________ Model ____________________

VIN ____________________ Engine type and size ____________________

PROCEDURE

NOTE: *Brake rotors should be refinished only under the following conditions:*

- If the thickness of the rotor is great enough to allow metal removal and still be above the minimum wear dimension.
- If it fails lateral runout or thickness variation checks.
- If there is noticeable brake pulsation.
- If there are heat spots or excessive scoring that can be removed by resurfacing.

1. Here are guidelines to adhere to when refinishing rotors:
   - Whenever you refinish a rotor, remove the least amount of metal possible to achieve the proper finish.
   - Never turn the rotor on one side of the vehicle without turning the rotor on the other side.
   - Equal amounts of metal should be cut off both surfaces of a rotor.
   - Do not refinish new rotors unless the measured runout exceeds specifications.
   - Clean any oil film off the rotor with brake cleaning solvent or alcohol and let the rotor air dry before installing it on the vehicle.

2. Describe the type of lathe you will be using for this job sheet.

____

____

3. Check the lathe for condition, as well as the attaching adapters, tool holders, vibration dampers, and cutting bits. Make sure the mounting adapters are clean and free of nicks. Always use sharp cutting tools or bits and use only replacement cutting bits recommended by the equipment manufacturer. The tip of the cutting bit should be slightly rounded. State the results of this check.

____

____

4. Mount the rotor onto the lathe. A one-piece rotor with bearing races in its hub mounts to the lathe arbor with tapered or spherical cones. A two-piece rotor removed from its hub is centered on the lathe arbor with a spring-loaded cone and clamped in place by two large cup-shaped adapters. ☐ Task completed

5. Remove all grease and dirt from the bearing races before mounting the rotor. Make sure the bearing races are not loose in the rotor. Use the appropriate cones and spacers to lock the rotor firmly to the arbor shaft. When mounting a two-piece rotor without the hub, clean all rust and corrosion from the hub area with emery cloth. Use the proper cones and spacers to mount the rotor to the arbor shaft. ☐ Task completed

6. After the rotor is on the lathe, install the vibration damper band on the outer diameter of the rotor to prevent the cutting bits from chattering during refinishing. ☐ Task completed

7. Before removing any metal from the rotor, verify that it is centered on the lathe arbor. ☐ Task completed

8. Make a small scratch on one surface of the rotor to check rotor mounting. Do this by backing the cutting assembly away from the rotor and turning the rotor through one complete revolution to be sure there is no interference with rotation. ☐ Task completed

9. Start the lathe and advance the cutting bit until it just touches the rotor surface near midpoint. ☐ Task completed

10. Let the cutting bit lightly scratch the rotor, approximately 0.001 inch (0.025 mm) deep. ☐ Task completed

11. Move the cutting bit away from the rotor and stop the lathe. If the scratch is all the way around the rotor, the rotor is centered and you can proceed with resurfacing ☐ Task completed

12. If the scratch appears as a crescent, either the rotor has a lot of runout or it is not centered on the arbor. In this case, loosen the arbor nut and rotate the rotor 180 degrees on the arbor; then retighten the nut. ☐ Task completed

13. Repeat steps 9 through 11 to make another scratch about 1/4 inch away from the first. ☐ Task completed

14. If the second scratch appears at the same location on the surface as the first, the rotor has significant runout, but it is properly centered on the lathe and you can proceed with machining. If the second scratch appears opposite the first on the rotor surface, remove the rotor from the lathe arbor and recheck the mounting. ☐ Task completed

15. To determine the approximate amount of metal to be removed, turn on the lathe and bring the cutting bit up against the rotating disc until a slight scratch is visible as you did to verify rotor centering. ☐ Task completed

16. Turn off the lathe and reset the depth-of-cut dial indicator to zero. Find the deepest groove on the face of the rotor and move the cutting bit to that point without changing its depth-of-cut setting. Now use the depth-of-cut dial to bottom the tip of the cutter in the deepest groove. The reading on the dial now equals, or is slightly less than, the amount of metal to be removed to eliminate all grooves in the rotor. ☐ Task completed

17. To make the series of refinishing cuts, begin by resetting the cutter position so the cutting bits again just touch the ungrooved surface of either side of the rotor. Then, zero the depth-of-cut indicators on the lathe. ☐ Task completed

18. Turn on the lathe and let the arbor reach full running speed. ☐ Task completed

19. Turn the depth-of-cut dials for both bits to set the first pass cut. Turning these dials moves the bits inward. The dial is calibrated in thousandths-of-an-inch increments. The first cut should only be a portion of the total anticipated depth of cut. ☐ Task completed

20. When the cutting depth is set for the first cut, activate the lathe to move the cutting bits across the surface of the rotor. After the first cut is completed, turn off the lathe and examine the rotor surface. Areas that have not yet been touched by the bits will be darker than those that have been touched. ☐ Task completed

21. If there are large patches of unfinished surface, make another cut of the same depth. When most of the surface has been refinished, make a shallow finishing cut at lower arbor speed. Repeat the slow finishing cut until the entire rotor surface has been refinished. ☐ Task completed

22. Make sure you do not cut the rotor thickness beyond its service limit. To ensure this, re-measure the refinished rotor with a micrometer to determine its minimum thickness and compare this measurement to the manufacturer's minimum refinished thickness specification. ☐ Task completed

23. To make sure the finish of the rotor is non-directional after machining, dress the rotor surfaces with a sanding disc power tool with 120- to 150-grit aluminum oxide sandpaper. Sand each rotor surface with moderate pressure for at least 60 seconds. ☐ Task completed

24. After the rotor has been sanded, clean each surface with hot water and detergent. Then dry the rotors thoroughly with clean paper towels. ☐ Task completed

25. Record any difficulties you had completing this job sheet.

____

____

Instructor's Comments

BRAKES JOB SHEET 21

Servicing a Rear Disc Caliper

Name ______________________ Station ______________ Date ______________

NATEF Correlation

This Job Sheet addresses the following NATEF task:

D.10. Adjust calipers with integrated parking brake system.

Objective

Upon completion of this job sheet, you will be able to adjust the calipers with the integrated parking brake system.

Tools and Materials

Appropriate screwdrivers	Service manual
Appropriate wrenches	Torque wrench
Brake fluid	Wire
C-clamps	Wood block

Protective Clothing

Goggles or safety glasses with side shields

Respirator

Describe the vehicle being worked on:

Year ________ Make ____________________ Model ____________________

VIN ____________________ Engine type and size ____________________

Describe general condition:

____

PROCEDURE

1. Remove the wheel. Screw one wheel lug back on to keep the rotor in place. ☐ Task completed
2. Loosen the parking brake cable at the equalizer, and remove the cable from the actuating lever. Remove the actuating lever return spring. ☐ Task completed
3. Hold the actuating lever in place and remove the actuating lever lock nut. Remove the actuating lever seal and antifriction washer. ☐ Task completed
4. Bottom the pistons with a C-clamp. Do not position the C-clamp on the actuator screw. ☐ Task completed
5. Disconnect the brake line. ☐ Task completed

6. Remove the caliper guide pins. ☐ Task completed
7. Lift and drain the caliper. ☐ Task completed
8. Push the two metal sleeves out of the inner caliper ears. ☐ Task completed
9. Remove the O-rings from the inner and outer ears. ☐ Task completed
10. Replace the lever and rotate it back and forth until the piston comes out of the bore. Remove the piston assembly and balance spring, lock nut, lever seal, and antifriction washer. ☐ Task completed
11. Push the actuating screw out of the caliper. ☐ Task completed
12. Clean and overhaul the caliper and all parts. Left- and right-hand parts are not interchangeable. ☐ Task completed
13. Assemble the new thrust screw with a new washer and seal into the piston. Install the spring and piston assembly into the caliper bore. ☐ Task completed
14. Bottom the piston with a special tool. Lubricate and install the antifriction washer, new lever seal, lever, and lock nut. ☐ Task completed
15. Install the lever away from the stop, rotate it forward, and hold it in place. Torque it to 25 ft-lb, and remove the special tool. ☐ Task completed
16. Rotate the lever back to the stop on the caliper. Install the return spring. ☐ Task completed
17. Use the special tool to seat the dust boot ring. ☐ Task completed
18. Lubricate and install the O-rings into the inner and outer ears. ☐ Task completed
19. Lubricate the outer surface of the sleeves and install them into the inner ear with the special tool. ☐ Task completed
20. Install the inner pad with a D-shaped tab, turning the piston if necessary. ☐ Task completed
21. Install the outer piston with the ears of the shoe over the two outer ears of the caliper and the bottom flange fitting. ☐ Task completed
22. Clean and coat the guide pins with assembly fluid. Slide the caliper into place so the caliper holes and the holes on the support bracket correspond. Push the pins under the inner pad ears, into the outer pad ears, and thread them into the support bracket threads. Torque the pins. ☐ Task completed
23. Connect the brake hose. ☐ Task completed
24. Bleed the system and pump the pedal to seat the pads. ☐ Task completed
25. Crimp the shoe ears with pliers. ☐ Task completed
26. Connect the parking brake cable to the actuating lever. ☐ Task completed
27. Operate the parking brake. If the actuating lever does not return to the stop, loosen the cable at the equalizer. If it still fails to return, replace the return spring or piston and screw assembly. ☐ Task completed

Problems Encountered

Instructor's Comments

BRAKES JOB SHEET 22

Installing a Tire and Wheel Assembly

Name ______________________ Station ______________ Date ______________

NATEF Correlation

This Job Sheet addresses the following NATEF task:

D.11. Install wheel, torque lug nuts, and make final checks and adjustments.

Objective

Upon completion of this job sheet, you will be able to install a wheel, torque lug nuts, and make final checks and adjustments.

Tools and Materials

Breaker bar

Hydraulic floor jack

Wheel chocks

Safety stands

Protective Clothing

Goggles or safety glasses with side shields

Describe the vehicle being worked on:

Year ________ Make ____________________ Model ____________________

VIN __________________ Tire size ______________ Wheel Diameter ______

PROCEDURE

1. Determine the proper lift points for the vehicle. Describe them here:

   ____

   ____

2. Place the pad of the hydraulic floor jack under one of the lift points for the left side of the front of the vehicle. Describe this location.

   ____

   ____

3. Raise the pad just enough to make contact with the vehicle. ☐ Task completed

4. Make sure the vehicle's transmission is placed in park or in a low gear (if there is a manual transmission) and apply the parking brake. ☐ Task completed

5. Place blocks or wheel chocks at the rear tires to prevent the vehicle from rolling. ☐ Task completed

6. Remove any hub or bolt caps to expose the lug nuts or bolts. ☐ Task completed

7. Refer to the service or owner's manual to determine the order the lugs ought to be loosened. If this is not available, make sure they are alternately loosened. Describe the order you will use to loosen and remove the lug nuts.

____

____

8. Get the correct size socket or wrench for the lug nuts. The size is: ________

9. Using a long breaker bar, loosen each of the lug nuts but do not remove them. ☐ Task completed

10. Raise the vehicle with the jack. Make sure it is high enough to remove the wheel assembly. Once it is in position, place a safety stand under the vehicle. Describe the location of the stand.

____

____

11. Lower the vehicle onto the stand and lower and remove the jack from the vehicle. ☐ Task completed

12. Remove the lug nuts and the tire and wheel assembly. ☐ Task completed

13. Before installing the wheel and tire assembly, make sure the tire is inflated to the recommended inflation. What is that inflation?

____

14. Move the assembly over to the vehicle. ☐ Task completed

15. Place the assembly onto the hub and start each of the lug nuts. ☐ Task completed

16. Hand tighten each lug nut so that the wheel is fully seated against the hub. ☐ Task completed

17. Put the hydraulic jack into position and raise the vehicle enough to remove the safety stands. ☐ Task completed

18. Lower the jack so the tire is seated on the shop floor. ☐ Task completed

19. Following the reverse pattern as used to loosen the lug nuts, tighten the lug nuts according to specifications. The order you used to tighten was:

____

____

The recommended torque spec is: ____________________

20. What happens if you over tighten the lug nuts?

____

____

____

21. Install the bolt or hub cover. ☐ Task completed

Instructor's Comments

BRAKES JOB SHEET 23

Remove and Install a Typical Hub and Rotor Assembly

Name ______________________ Station ______________ Date ______________

NATEF Correlation

This Job Sheet addresses the following NATEF task:

D.8. Remove and reinstall rotor.

Objective

Upon completion of this job sheet, you will be able to correctly remove and replace a rotor.

Tools and Materials

Appropriate screwdrivers
Appropriate wrenches
Service manual
Wire

Protective Clothing

Goggles or safety glasses with side shields

Describe the vehicle being worked on:

Year ________ Make ____________________ Model ____________________

VIN ____________________ Engine type and size ____________________

Describe general condition:

____

PROCEDURE (REMOVAL)

1. Remove the wheel and tire assembly. A disc brake rotor cannot be removed with the wheel because the caliper straddles the rotor. ☐ Task completed
2. Remove the caliper assembly, but do not disconnect the brake hose unless the caliper is to be removed for service. If the caliper does not require servicing, suspend it from a wire hook or loop it on the front suspension to avoid strain on the hose or damage to the crossover line or other hydraulic connections. ☐ Task completed
3. If the rotor is integral with the wheel bearing hub, remove the grease cap from the hub and the cotter pin, adjusting nut, and thrust washer from the spindle. ☐ Task completed
4. Remove the outer wheel bearing cone and roller assembly from the hub. ☐ Task completed
5. Remove the hub and rotor assembly from the spindle. ☐ Task completed
6. If the splash shield has been damaged, replace it by removing the bolts that attach it to the knuckle. ☐ Task completed

PROCEDURE (REINSTALLATION)

1. Before installing the hub and rotor, make sure the inner and outer bearings are not damaged and that they are properly greased. Install a new grease seal. ☐ Task completed
2. Place the hub and rotor assembly over the spindle and push it in as far as it will go. ☐ Task completed
3. Install the outer bearing over the spindle followed by a thrust washer and nut. ☐ Task completed
4. Tighten the bearing adjusting nut while rotating the hub to the specified torque, then back it off and adjust it to the proper torque as given in the service manual. ☐ Task completed

   First torque specification ______________________________

   Second torque specification ______________________________
5. Install the lock washer, cotter pin, and grease cup. ☐ Task completed
6. Install the caliper over the rotor. ☐ Task completed
7. Install the wheel and tire. Road test the vehicle. ☐ Task completed

Problems Encountered

____

____

____

Instructor's Comments

____

____

____

BRAKES JOB SHEET 24

Vacuum Booster Testing and Diagnosis

Name ______________________________ Station __________________ Date __________________

NATEF Correlation

This Job Sheet addresses the following NATEF task:

E.1. Test pedal free travel with and without engine running; check power assist operation.

Objective

Upon completion of this job sheet, you will be able to test pedal free travel with and without the engine running and check power assist operation.

Tools and Materials

12-inch rule

Vacuum gauge

Protective Clothing

Goggles or safety glasses with side shields

Describe the vehicle being worked on:

Year __________ Make __________________________ Model __________________________

VIN __________________________ Engine type and size __________________________

PROCEDURE

1. Pump the brake pedal with the engine off to exhaust vacuum in the booster. ☐ Task completed
2. Place a ruler against the car floor, in line with the arc of pedal travel. ☐ Task completed
3. Press the pedal by hand and measure the amount of travel before looseness in the linkage, or freeplay, is taken up. ☐ Task completed
4. Measure at the top or bottom of the pedal, whichever provides the most accurate view. What was the measurement?

   ____

5. Start the engine. Place the ruler against the car floor, in line with the arc of pedal travel. ☐ Task completed
6. Press the pedal by hand and measure at the same spot as you did in step 4. What was the measurement?

   ____

7. The pedal should have moved slightly closer to the floor, but depressing the pedal should require less effort. What did you observe and what does this indicate?

____

____

8. If the freeplay did not change, as it should have, check engine vacuum before checking the vacuum booster system. Connect a vacuum gauge to the engine's intake manifold and measure vacuum while the engine is idling. State what the reading was and what is indicated by it.

____

9. With the engine running and the brakes applied, listen for a hiss. A steady hiss when the brake is held down indicates a leak in the booster. What did you observe?

____

____

10. If there seems to be less than normal power assist when depressing the brake pedal, check the vacuum booster's check valve. What were the results of this and what did you use to check the valve?

____

____

11. Check for a restricted vacuum hose by disconnecting it from the booster with the engine running. If the engine doesn't stumble and almost stall, the hose is probably restricted. What did you observe?

____

____

Instructor's Comments

____

____

____

BRAKES JOB SHEET 25

Perform a Power Vacuum Brake Booster Test

Name ______________________ Station ______________ Date ______________

NATEF Correlation

This Job Sheet addresses the following NATEF tasks:

E.2. Check vacuum supply (manifold or auxiliary pump) to vacuum-type power booster.

E.3. Inspect the vacuum-type power booster unit for vacuum leaks; inspect the check valve for proper operation; determine necessary action.

Objective

Upon completion of this job sheet, you will be able to check vacuum supply (manifold or auxiliary pump) to vacuum-type power booster. You will also be able to inspect the vacuum-type power booster unit for vacuum leaks and inspect the check valve for proper operation.

Tools and Materials

Appropriate wrenches and screwdrivers
Fender covers
Safety jacks or lift
Service manual
Vacuum gauge

Protective Clothing

Goggles or safety glasses with side shields

Describe the vehicle being worked on:

Year ________ Make ____________________ Model ____________________

VIN ____________________ Engine type and size ____________________

PROCEDURE (BASIC OPERATIONAL TEST)

1. Place fender covers over the fenders. With the engine off, pump the brake pedal numerous times to be sure any residual vacuum is exhausted from the booster unit. ☐ Task completed

2. Hold firm pressure on the brake pedal and start the engine. If the system is operating properly, you should feel the pedal move downward a short distance, then stop. Only a small amount of pressure should be needed to hold the pedal down. If you do not get these results, proceed to the other tests. Record your results on the Report Sheet for Performing Power Vacuum Brake Test. ☐ Task completed

PROCEDURE (VACUUM SUPPLY TEST)

1. With the engine idling, attach the vacuum gauge to the intake manifold port. A low reading indicates an engine problem. Record the results on the Report Sheet for Performing Power Vacuum Brake Test. ☐ Task completed
2. Disconnect the vacuum tube or hose that runs from the intake manifold to the booster unit, and place your thumb over it (do this quickly or the engine might stall). If you do not feel a strong vacuum at the end of the tube or hose, shut off the engine, remove the tube or hose, and see if it is collapsed, crimped, or clogged. Replace it if necessary. ☐ Task completed

PROCEDURE (VACUUM CHECK VALVE TEST)

1. Shut the engine off and wait five minutes. ☐ Task completed
2. Apply the brakes. There should be power assist on at least one pedal stroke. Record the results on the Report Sheet for Performing Power Vacuum Brake Test. ☐ Task completed
3. If no power assist is felt, remove the check valve from the booster unit. ☐ Task completed
4. Test the check valve by blowing into the intake manifold end of the valve. There should be a complete blockage of airflow. Alternatively, apply vacuum to the booster unit end of the valve. Vacuum should be blocked. ☐ Task completed
5. If the vacuum is not blocked, replace the check valve. ☐ Task completed

PROCEDURE (BRAKE DRAG TEST)

1. With the wheels raised off the floor, pump the brake pedal to exhaust the vacuum from the booster. ☐ Task completed

   WARNING: *Support the vehicle on jack stands or a hoist.*
2. Turn the front wheels by hand and note amount of drag that is present. ☐ Task completed
3. Start the engine and allow it to run for one minute. Then shut it off. ☐ Task completed
4. Turn the front wheels by hand again. If the drag has increased, the booster control valve is faulty. Record the results on the Report Sheet for Performing Power Vacuum Brake Test. ☐ Task completed

PROCEDURE (FLUID LOSS TEST)

1. If the master cylinder reservoir level has fallen noticeably, but there is no sign of an external brake fluid leak, remove the vacuum tube or hose that runs between the intake manifold and booster unit, and inspect it carefully for evidence of brake fluid. If any fluid is found, the master cylinder secondary seal is leaking. Record the results on the Report Sheet for Performing Vacuum Brake Test. ☐ Task completed

Problems Encountered

____

____

____

Instructor's Comments

____

____

____

Name ______________________ Station ______________ Date ______________

REPORT SHEET FOR PERFORMING POWER VACUUM BRAKE TEST		
	Yes	*No*
1. Pedal moved down when engine started		
2. Engine vacuum reading		
	Yes	*No*
3. Vacuum present at booster		
Hose condition		
	Yes	*No*
4. Power assist with engine off		
	OK	*Replace*
5. Check valve test		
	Yes	*No*
6. Drag increase		
7. Fluid loss		
Conclusions and Recommendations ______________________________		

BRAKES JOB SHEET 26

Checking a Hydro-Boost System

Name ______________________________ Station __________________ Date __________________

NATEF Correlation

This Job Sheet addresses the following NATEF task:

E.4. Inspect and test hydro-boost system and accumulator for leaks and proper operation; determine necessary action.

Objective

Upon completion of this job sheet, you will be able to inspect and test the hydro-boost system and accumulator for leaks and proper operation.

Tools and Materials

Clean rags
Power steering pressure tester
Flare nut wrenches
Hand tools
Service manual

Protective Clothing

Goggles or safety glasses with side shields

Describe the vehicle being worked on:

Year __________ Make __________________________ Model __________________________

VIN ______________________ Engine type and size __________________________

PROCEDURE

1. Check the basic operation of the engine and power steering system. If both the brake applications and steering require more than normal effort, the cause is probably related to fluid pressure and delivery in the power steering system. Summarize your findings.

____

____

2. Inspect fluid level in the power steering pump. Summarize your findings.

____

____

3. Inspect the condition of the power steering fluid. Summarize your findings.

____

____

4. Inspect the condition of the power steering pump drive belt. Check the tension of the drive belt and make sure the belt is installed properly and correctly positioned on the belt tensioner. Summarize your findings.

____

____

5. Inspect all hoses and steel lines in both the Hydro-Boost system and the power steering system for leakage. Summarize your findings.

____

____

6. Check the brake fluid level in the master cylinder and add fluid if necessary. Summarize your findings.

____

____

7. Check the pressure of the power steering system. Compare your readings with the specifications. Summarize your findings.

____

____

8. Check the Hydro-Boost system for signs of leakage. Summarize your findings.

____

____

9. With the engine off, pump the brake pedal repeatedly to bleed off the residual hydraulic pressure stored in the accumulator. ☐ Task completed

10. Hold firm pressure on the brake pedal and start the engine. The brake pedal should move downward, then push up against your foot. What happened?

____

____

11. Rotate the steering wheel until it stops and hold it in that position for no more than 5 seconds. ☐ Task completed

12. Return the steering wheel to the center position and shut off the engine. ☐ Task completed

13. Pump the brake pedal. You should feel two or three power-assisted strokes. Did you? What does this indicate?

____

____

14. Now repeat steps 10 and 11. This will pressurize the accumulator. ☐ Task completed

15. Turn the engine off and wait 1 hour; then pump the brake pedal. There should be one or two power-assisted strokes. Summarize what you found.

____

____

16. What are your service recommendations?

____

____

____

17. If the booster must be replaced, turn the engine off and pump the brake pedal several times to exhaust accumulator pressure. ☐ Task completed

18. Disconnect the master cylinder from the booster, but leave the service brake hydraulic lines connected to the master cylinder. ☐ Task completed

19. Carefully lay the master cylinder aside, being careful not to kink or bend the steel tubing. Support the master cylinder from a secure point on the vehicle with wire or rope. Do not support the master cylinder on the brake lines. ☐ Task completed

20. Disconnect the hydraulic hoses from the booster ports. Plug all tubes and the booster ports to prevent fluid loss and system contamination. ☐ Task completed

21. Detach the pedal pushrod from the brake pedal. Remove the nuts and bolts from the booster support bracket and remove the booster from the vehicle. ☐ Task completed

22. To reinstall the Hydro-Boost booster, connect the hydraulic hoses. ☐ Task completed

23. Install the master cylinder. ☐ Task completed

24. Fill the power steering pump reservoir to the full mark and allow it to sit undisturbed for several minutes. ☐ Task completed

25. Start the engine and run it for approximately one minute. ☐ Task completed

26. Stop the engine and recheck the fluid level. ☐ Task completed

27. Raise the front of the vehicle on a hoist or safety stands. ☐ Task completed

28. Turn the wheels from lock to lock. Check and add fluid if needed. ☐ Task completed

29. Lower the vehicle and start the engine. ☐ Task completed

30. Apply the brake pedal several times while turning the steering wheel from lock to lock. ☐ Task completed

31. Turn off the engine and pump the brake pedal five or six times. ☐ Task completed

32. Recheck the fluid level. If the steering fluid is extremely foamy, allow the vehicle to stand for at least 10 minutes with the engine off. ☐ Task completed

Instructor's Comments

____

____

____

BRAKES JOB SHEET 27

Diagnosing Wheel and Tire Problems

Name ______________________________ Station __________________ Date __________________

NATEF Correlation

This Job Sheet addresses the following NATEF task:

F.1. Diagnose wheel bearing noises, wheel shimmy, and vibration concerns; determine necessary action.

Objective

Upon completion of this job sheet, you will be able to diagnose wheel bearing noises, wheel shimmy, and vibration concerns.

Tools and Materials

Tread wear gauge

Protective Clothing

Goggles or safety glasses with side shields

Describe the vehicle being worked on:

Year __________ Make __________________________ Model __________________________

VIN _______________________ Tire manufacturer and size ___________________________

PROCEDURE

1. Drive the vehicle at a variety of speeds and conditions and note all noises and handling problems that occurred during the road test.

   ____

   ____

   ____

2. Check the suspension, steering, driveline, and brake system to make sure they are not the cause of the problems. Describe the results of this check.

   ____

   ____

3. Check the wear patterns of the tires to help eliminate the suspension and steering systems as causes of the problem. Describe the wear pattern of the tires.

   ____

   ____

4. Check the wear of the tires with a tread wear gauge or by looking for tread bars on the tires. Describe the wear of each tire.

____

____

5. Check the inflation of all of the tires and compare the pressure to specifications. What did you read and what are the specifications?

____

____

6. If the results from the above all indicate that the tires; suspension, steering, and brake systems; and the driveline are in good condition and a noise was evident during the test drive, suspect the tire design as the cause of the problem. Describe the tread design of the tires. Do all of the tires have the same design?

____

____

7. What can you conclude about the source of the abnormal noise?

____

____

8. Describe any vibration problems felt during the road test. If the tires felt as if they were bouncing along, the problem is typically called wheel tramp. If the tires seemed to vibrate while turning, the problem is called shimmy.

____

____

____

____

9. If wheel tramp is the problem, the tire and wheel assemblies need to be checked for static imbalance. If wheel shimmy is the problem, check the tire and wheel assemblies for dynamic imbalance. Describe the results of this check.

____

____

10. If the tire and wheel assembly are both statically and dynamically balanced, check the wheel hub or axle for excessive runout. ☐ Task completed

11. If runout is not the problem, check for loose, worn, or damaged wheel bearings. ☐ Task completed

Instructor's Comments

____

____

____

BRAKES JOB SHEET 28

Remove and Install Front-Wheel Bearings on a RWD Vehicle

Name ______________________________ Station __________________ Date __________________

NATEF Correlation

This Job Sheet addresses the following NATEF tasks:

F.2. Remove, clean, inspect, repack, and install wheel bearings and replace seals; install hub and adjust wheel bearings.

F.7. Replace wheel bearing and race.

Objective

Upon completion of this job sheet, you will be able to remove, clean, inspect, repack, and install wheel bearings and replace seals. You will also be able to install the hub and adjust the wheel bearings, as well as replace the wheel bearing and race.

Tools and Materials

Ball peen hammer
Bearing repacker, optional
Clean paper
Drift
Grease
Hoist or jack stands
Installation tools, such as arbor press, press-fitting tools with an outside diameter approximately 0.010″ smaller than bore size, soft striking mallet, and wood block and hammer—if no other tools are available
Lint-free cloth
New bearings
New seals
Pliers or screwdriver
Service manual
Torque wrench
Wire, if needed
Wrenches

Protective Clothing

Goggles or safety glasses with side shields

Describe the vehicle being worked on:

Year __________ Make ________________________ Model ________________________

VIN ________________________ Engine type and size ________________________

Describe general condition:

____

PROCEDURE (REMOVAL)

1. Raise the front end of the vehicle on a hoist or safely support it on jack stands. Do not support the vehicle on only a bumper jack. Remove the hub cap or wheel cover. ☐ Task completed
2. Use a wrench to remove the wheel lug nuts. If the vehicle is equipped with disc brakes, loosen and remove the brake caliper mounting bolts. Support the caliper while disconnected on the lower A-frame or suspended by a wire loop. ☐ Task completed ☐ Not applicable
3. Use pliers or a screwdriver to remove the dust cover. Remove the cotter pin, and remove the adjusting nut. ☐ Task completed
4. Jerk the rotor or drum assembly to loosen the washer and outer wheel bearing. If this step is not done easily, the drum brakes might have to be backed off. ☐ Task completed
5. Remove the outer wheel bearing. Then pull the drum or rotor assembly straight off the spindle. Make sure the inner bearing or seal does not drag on the spindle threads. ☐ Task completed
6. With the seal side down, lay the rotor or drum on the floor. Place the drive against the inner race of the bearing cone. Carefully tap out the old seal and inner bearing. ☐ Task completed

   WARNING: *Wear eye protection whenever using a hammer and drift punch.*

7. Record the part number of the seal on the Report Sheet for Removal and Installation of Front Wheel Bearings and Seals, to aid in selecting the correct placement. Discard the old seal. ☐ Task completed
8. Clean and inspect the old bearing thoroughly. ☐ Task completed

   WARNING: *Do not use air pressure to spin the bearing during cleaning. A lack of lubrication can cause the bearing to explode, resulting in serious injury.*

9. Inspect the bearing to determine if it can be reused. Record the results on the Report Sheet for Removal and Installation of Front Wheel Bearings and Seals. If it must be replaced, record the part number on the report sheet. ☐ Task completed

PROCEDURE (INSTALLATION)

1. Match the part numbers to make sure the new seal is correct for the application. ☐ Task completed
2. By hand or with a bearing repacker, force grease through the cage and rollers or balls and on all surfaces of the bearing. ☐ Task completed
3. Place the inner side of the drum or rotor face up. Use drivers to drive the new cup into the hub. ☐ Task completed

4. Coat the hub cavity with the same wheel bearing grease to the depth of the bearing cup's smallest diameter. Apply a light coating of grease to the spindle. ☐ Task completed

5. Place the inner bearing on the hub. Lightly coat the lip of the new seal with the same grease. Slide the seal onto the proper installation tool. Make sure the seal fits over the tool's adapter and the sealing lip points toward the bearing. ☐ Task completed

6. Position the seal so it starts squarely in the hub without cocking. Tap the tool until it bottoms out. (When the sound of the striking mallet changes, the seal will be fully seated in the hub.) ☐ Task completed

7. If the installation tool is unavailable, use a wood block and hammer to drive the seal. (Never hammer directly on seal.) ☐ Task completed

8. Locate the lug nut and wheel bearing adjusting nut torque specifications in the service manual. Record the specifications on the Report Sheet for Removal and Installation of Front Wheel Bearings and Seals. Using these specifications and service manual procedures, install the hub assembly to the spindle. ☐ Task completed

Problems Encountered

____

____

____

Instructor's Comments

____

____

____

Name ______________________ Station ______________ Date ______________

REPORT SHEET FOR REMOVAL AND INSTALLATION OF FRONT WHEEL BEARINGS AND SEALS		
Seal part number		
Bearing part number		
	Serviceable	*Nonserviceable*
Bearing inspection		
Cup		
Rollers		
Inner race		
Cage		
Overall condition		
Lug nut torque specification		
Bearing adjustment nut torque		
Initial torque		
Number of turns backed off		
Final torque		

Conclusions and Recommendations ____

____

____

BRAKES JOB SHEET 29

Adjust and Replace a Parking Brake Cable

Name ______________________ Station ______________ Date ______________

NATEF Correlation

This Job Sheet addresses the following NATEF task:

F.3. Check parking brake cables and components for wear, rusting, binding, and corrosion; clean, lubricate, and replace as needed.

Objective

Upon completion of this job sheet, you will be able to check parking brake cables and components for wear, rusting, binding, and corrosion; and clean. You will also be able to lubricate and replace the components as necessary.

Tools and Materials

Pliers
Service manual
Screwdriver assortment
Wrench assortment

Protective Clothing

Goggles or safety glasses with side shields
Respirator

Describe the vehicle being worked on:

Year ________ Make __________________ Model __________________

VIN __________________ Engine type and size __________________

PROCEDURE (REAR CABLE REMOVAL)

1. Raise the vehicle to a convenient working height on a hoist. Remove the rear wheels and disconnect the brake cable. ☐ Task completed
2. Detach the retaining clip from the brake cable retainer bracket. Remove the brake drum (from the rear axle), brake shoe return springs, and brake shoe retaining springs. ☐ Task completed

   WARNING: *After removing the brake drum, install the asbestos removal vacuum. Follow the manufacturer's instructions for the removal of asbestos from the brake components.*

3. Remove the brake shoe strut and spring from the brake support plate. ☐ Task completed
4. Disconnect the brake cable from the operating level. ☐ Task completed
5. Compress the retainers on the end of the brake cable housing and remove the cable from the support plate. ☐ Task completed

PROCEDURE (REAR CABLE INSTALLATION)

1. Insert the brake cable and housing into the brake support plate, making sure that the housing retainers lock the housing firmly into place. ☐ Task completed
2. Hold the brake shoes in place on the support plate. Engage the brake cable into the brake shoe operating lever. ☐ Task completed
3. Install the parking brake strut and spring, brake shoe retaining springs, brake shoe return springs, and brake drum and wheel. ☐ Task completed
4. Insert the brake cable and housing into the bracket and install the retaining clip. ☐ Task completed
5. Insert the brake cable into the equalizer. ☐ Task completed
6. Adjust the rear service brakes and parking brake cable. ☐ Task completed

PROCEDURE (PARKING BRAKE ADJUSTMENT—REAR DRUM)

1. Raise the vehicle and support it with jack stands placed under the suspension. Set the transmission shift lever in the "neutral" position. ☐ Task completed
2. Release the parking brake operating lever and loosen the cable adjusting nut to make sure the cable is slack. Clean the cable threads with a wire brush, and lubricate them to prevent rusting. ☐ Task completed
3. Tighten the cable adjusting nut at the equalizer until a slight drag is felt while rotating the rear wheels. ☐ Task completed
4. Loosen the cable adjusting nut until both rear wheels can be rotated freely. Back off the cable adjusting nut two full turns. There should be no drag at the rear wheels. ☐ Task completed
5. Make a final operational check of the parking brake. Then lower the vehicle. ☐ Task completed

PROCEDURE (FRONT CABLE REMOVAL)

1. Using a screwdriver, force the cable housing and attaching clip forward out of the body cross-member. ☐ Task completed
2. Fold back the left edge of the floor covering to get at and pry out either the rubber grommet or the rubber cable cover from the floor pan (depending on car model). Remove the floor pan clip. ☐ Task completed
3. Engage the parking brake and work the brake cable up and out of the clevis linkage or guide bracket. ☐ Task completed
4. Using a screwdriver, force the cable assembly and clip down and out of the pedal assembly bracket. If necessary, attach a substantial wire about 50 inches long to the lower end of the cable housing (to be used as a guide for cable installation). ☐ Task completed

5. Work the cable and housing assembly up through the floor pan. If used, disconnect the guide wire, but leave it in place for new front cable installation. ☐ Task completed

PROCEDURE (FRONT CABLE INSTALLATION)

1. Pull the guide wire (if used) connected to the new cable or the cable itself through the floor pan. ☐ Task completed
2. Insert the retainer into the hole in the bottom of the parking brake pedal assembly bracket. ☐ Task completed
3. Insert the cable through the hole and cable end fitting into the clevis linkage or guide bracket. ☐ Task completed
4. Force the upper cable housing into the retainer until it is firmly seated against the pedal assembly bracket. ☐ Task completed
5. Insert the retainer into the hole in the cross-member or bracket. ☐ Task completed
6. Insert the cable through the hole in the cross-member and force the cable housing into the retainer until it is firmly seated. ☐ Task completed
7. Where applicable, push the cable seal into the hole in the dash panel with enough force to seat the locking tabs. Otherwise, reinstall the rubber cable cover and floor pan clip. ☐ Task completed
8. Attach the cable (through bracket, if necessary) to the connector. ☐ Task completed
9. Adjust the rear service brakes and parking brake. ☐ Task completed
10. Apply the parking brake several times and test for free wheel rotation each time you release it. ☐ Task completed

Problems Encountered

____

____

____

Instructor's Comments

____

____

____

BRAKES JOB SHEET 30

Parking Brake Tests

Name ______________________ Station ______________ Date ______________

NATEF Correlation

This Job Sheet addresses the following NATEF task:

F.4. Check parking brake operation; determine necessary action.

Objective

Upon completion of this job sheet, you will be able to correctly check parking brake operations.

Tools and Materials

Basic hand tools

Protective Clothing

Goggles or safety glasses with side shields

Describe the vehicle being worked on:

Year ________ Make __________________ Model __________________

VIN __________________ Engine type and size __________________

PROCEDURE

1. Check the regular brake system for proper operation. What are the results of your check?

____

____

2. Inspect and test the control pedal or lever and the linkage for the parking brake. What are the results of your check?

____

____

3. Check the ability of the parking brake to hold the vehicle stationary. Use the guidelines given in the service manual for doing this. What are the results of your check?

____

____

4. If the parking brake lever or pedal must be applied to the full limit of its travel to engage the parking brakes, excessive clearance or slack exists somewhere in the system. To isolate the looseness in the system, press the service brake pedal and note its travel. If brake pedal travel seems excessive, the rear drum service brakes may need adjustment. Check and adjust the service brakes before adjusting the parking brake linkage. If the service brake pedal is firm and has normal travel, the parking brakes should be fully applied when the lever or pedal moves one-third to two-thirds of its travel. What are the results of your check?

____

____

5. If the parking brake lever or pedal travels more than specified, the linkage adjustment is too loose. ☐ Task completed

6. If the parking brake lever or pedal travels less than specified, the linkage adjustment is too tight. ☐ Task completed

7. If the lever or pedal travel is within the specified range, raise the vehicle on a hoist with the parking brakes released. Rotate the rear wheels by hand and check for brake drag. Have an assistant operate the parking brake control while you check the movement of the cables and linkage. The pedal or lever should apply smoothly and return to its released position. The parking brake cables should move smoothly without any binding or slack. What are the results of your check?

____

____

8. Check the protective conduit on the parking brake cables. What are the results of your check?

____

____

9. Inspect the cables for broken strands, corrosion, and kinks or bends. Damaged cables and linkage must be replaced. What are the results of your check?

____

____

10. Clean and lubricate the parking brake and non-coated metal cables with a brake lubricant. The parking brake cables on some vehicles are coated with a plastic material. This plastic coating helps the cables slide smoothly against the nylon seals inside the conduit end fittings. It also protects the cable against corrosion damage. Plastic-coated cables do not need periodic lubrication. ☐ Task completed

11. A small amount of drag is normal for rear drum brakes, but heavy drag usually indicates an over-adjusted parking brake mechanism. To determine if the misadjustment is in the linkage or in the caliper self-adjusters, disconnect the cables at a convenient point so that all tension is removed from the caliper levers. If brake drag is reduced, inspect the external linkage and lubricate, adjust, or repair it as necessary. If brake drag is still excessive with the parking brake linkage disconnected, check the caliper levers to be sure that they are returning fully against their stops. If the caliper levers are operating correctly and brake drag remains excessive, the calipers must be repaired or replaced. What are the results of your check?

___

___

___

Instructor's Comments

___

___

___

BRAKES JOB SHEET 31

ABS Diagnosis

Name ______________________ Station ______________ Date ______________

NATEF Correlation

This Job Sheet addresses the following NATEF tasks:

G.1. Identify and inspect and test anti-lock brake system (ABS) components; determine necessary action.

G.2. Diagnose poor stopping, wheel lock-up, abnormal pedal feel or pulsation, and noise concerns caused by the anti-lock brake system (ABS); determine necessary action.

Objective

Upon completion of this job sheet, you will be able to inspect and test anti-lock brake system components. You will also be able to diagnose poor stopping, wheel lock-up, abnormal pedal feel or pulsation, and noise concerns caused by the anti-lock brake system (ABS).

Tools and Materials

Basic hand tools

Service manual

Protective Clothing

Goggles or safety glasses with side shields

Describe the vehicle being worked on:

Year __________ Make ______________________ Model ______________________

VIN ______________________ Engine type and size ______________________

PROCEDURE

NOTE: *Always follow the vehicle manufacturer's procedures when diagnosing antilock brake systems. In general, ABS diagnostics requires three to five different types of testing that must be performed in the specified order listed in the service manual. Types of testing may include the following: pre-diagnostic inspections and test drive; warning light symptom troubleshooting; on-board ABS control module testing (trouble code reading); and individual trouble code or component troubleshooting. Procedures for conducting the system's self-test and component testing are included in another job sheet.*

1. Place the ignition switch in the START position while observing both the red brake warning light and the amber ABS indicator lights. Both lights should turn on. Start the vehicle. The red brake system light should quickly turn off. Describe what happened.

____

____

2. With the ignition switch in the RUN position, the antilock brake control module will perform a preliminary self-check on the antilock electrical system. The self-check takes three to six seconds, during which time the amber ABS light should remain on. Once the self-check is complete, the ABS light should turn off. If any malfunction is detected during this test, the ABS light will remain on and the ABS is shut down. Describe what happened.

____

____

3. Use the following as guidelines for determining what the brake warning lights are telling you. After comparing your test results with these guidelines, explain what was indicated by the light display you observed.
   - If the ECM detects a problem with the system, the amber ABS indicator lamp will either flash or light continuously to alert the driver to the problem. In some systems, a flashing ABS indicator lamp indicates that the control unit detected a problem but has not suspended ABS operation. However, a flashing ABS indicator lamp is a signal that repairs must be made to the system as soon as possible.
   - A solid ABS indicator lamp indicates that a problem has been detected that affects the operation of ABS. No antilock braking will be available. Normal, non-antilock brake performance will remain.
   - The red BRAKE warning lamp will be illuminated when the brake fluid level is low, the parking brake switch is closed, the bulb test switch section of the ignition switch is closed, or when certain ABS trouble codes are set.

____

____

4. Conduct a thorough visual inspection of all ABS and brake components. Also check for worn or damaged wheel bearings and check the condition of the tires and wheels. Look for signs that the vehicle may have poor wheel alignment. Describe what you found and what needs to be done before further diagnosis of the ABS. Remember that faulty base brake system components may cause the ABS system to shut down. Don't condemn the ABS system too quickly.

____

____

5. After the visual inspection is completed, test drive the vehicle to evaluate the performance of the entire brake system. Begin the test drive with a feel of the brake pedal while the vehicle is sitting still. ☐ Task completed

6. Then accelerate to a speed of about 20 mph. Bring the vehicle to a stop using normal braking procedures. Look for any signs of swerving or improper operation. Describe what happened.

____

____

7. Next, accelerate the vehicle to about 25 mph and apply the brakes with firm and constant pressure. You should feel the pedal pulsate if the antilock brake system is working properly. Describe what happened.

____

____

8. During the test drive, both brake warning lights should remain off. If either of the lights turns on, take note of the condition that may have caused it. After you have stopped the vehicle, place the gear selector into PARK or NEUTRAL and observe the warning lights. They should both be off. Describe what happened.

____

____

9. Summarize the condition of the ABS on this vehicle.

____

____

____

____

Instructor's Comments

____

____

____

BRAKES JOB SHEET 32

Perform ABS Tests with a Scan Tool

Name ______________________________ Station __________________ Date __________________

NATEF Correlation

This Job Sheet addresses the following NATEF task:

G.3. Diagnose anti-lock brake system (ABS) electronic control(s) and components using self-diagnosis and/or recommended test equipment; determine necessary action.

Objective

Upon completion of this job sheet, you will be able to diagnose anti-lock brake system (ABS) electronic controls and components using self-diagnosis and/or recommended test equipment.

Tools and Equipment

Scan tool

Service manual

Protective Clothing

Goggles or safety glasses with side shields

Describe the vehicle being worked on:

Year __________ Make ________________________ Model ________________________

VIN ________________________ Engine type and size ________________________

Describe the type of ABS found on the vehicle:

____

PROCEDURE

1. Turn the ignition on and check the operation of the ABS warning lights. If they do not light for a short time after the ignition is turned on, proceed to diagnose that problem before continuing with this procedure. What happened and what is indicated by the action of the lights?

   ____

   ____

   ____

2. Conduct a visual inspection:

   a. Check the master cylinder fluid level. Record your findings:

      ____

      ____

b. Inspect all brake hoses, lines, and fittings for signs of damage, deterioration, and leakage. Inspect the hydraulic modulator unit for any leaks or wiring damage. Record your findings:

____

____

c. Inspect the brake components at all four wheels. Make sure that no brake drag exists and that all brakes react normally when they are applied. Record your findings:

____

____

d. Inspect for worn or damaged wheel bearings that may allow a wheel to wobble. Record your findings:

____

____

e. Check the alignment and operation of the outer CV joints. Record your findings:

____

____

f. Make sure the tires meet the legal tread depth requirements and that they are the correct size. Record your findings:

____

____

g. Inspect all electrical connections for signs of corrosion, damage, fraying, and disconnection. Record your findings:

____

____

h. Inspect the wheel speed sensors and their wiring. Check the air gaps between the sensor and ring; make sure these gaps are within the specified range. Also check the mounting of the sensors and the condition of the toothed ring and wiring to the sensor. Record your findings:

____

____

3. Turn the ignition off. ☐ Task completed

4. Attach the scan tool test adapter—Select the adapter for the vehicle to be tested and attach it to the data cable using the two captive screws. ☐ Task completed

5. Attach the power cable to the test adapter and connect the power cable to the vehicle. (Skip this step for OBD-II adapter connection.) ☐ Task completed

6. Connect the test adapter to the vehicle's DLC. Make sure it is installed securely. ☐ Task completed

7. Turn the ignition on before selecting a test from a scan tool test menu. ☐ Task completed

8. Following the procedures given with the scan tool and in the service manual, retrieve the trouble codes and conduct any available tests on the system. Describe what codes were retrieved and what was indicated by them:

____

____

____

9. What are your recommendations for service as a result of this self-test?

____

____

____

10. Once all system malfunctions have been corrected, clear the ABS's DTCs. Codes cannot be erased until all codes have been retrieved, all faults have been corrected, and the vehicle has been driven above a set speed (usually 18 to 25 mph). It may be necessary to disconnect a fuse for several seconds to clear the codes on some systems. ☐ Task completed

11. After service work is performed on the ABS system, repeat the previous test procedure to confirm that all codes have been erased. ☐ Task completed

Problems Encountered

____

____

____

Instructor's Comments

____

____

____

BRAKES JOB SHEET 33

Relieving Accumulator Pressure

Name ______________________________ Station __________________ Date __________________

NATEF Correlation

This Job Sheet addresses the following NATEF task:

G.4. Depressurize high-pressure components of the anti-lock brake system (ABS).

Objective

Upon completion of this job sheet, you will be able to depressurize high-pressure components of an anti-lock brake system.

Tools and Materials

Bleeder T-wrench
Syringe
Drain hose and container

Protective Clothing

Goggles or safety glasses with side shields

Describe the vehicle being worked on:

Year __________ Make _________________________ Model _________________________

VIN _________________________ Engine type and size _________________________

PROCEDURE

NOTE: *Some brake services require that brake tubing or hoses be disconnected. Many ABS systems use hydraulic pressures as high as 2,800 psi and an accumulator to store this pressurized fluid. Before disconnecting any lines or fittings, the accumulator must be fully depressurized.*

1. Turn the ignition switch to the OFF position. ☐ Task completed
2. Pump the brake pedal between 25 and 50 times. ☐ Task completed
3. The pedal should be noticeably harder. Describe how the pedal feels.

   ____

   ____

4. Some manufacturers require the use of a special bleeder T-wrench to relieve the pressure in the accumulator and associated lines. Drain the brake fluid from the master cylinder and modulator reservoir thoroughly. The brake fluid can be removed from the reservoir by pulling it out through the top of the reservoir with a syringe. To drain the master cylinder, loosen the bleeder screw, install a drain hose onto the bleeder screw, and pump the brake pedal. ☐ Task completed
5. Remove the cap on the bleeder at the top of the ABS power unit. ☐ Task completed
6. Install the T-wrench on the bleeder screw and slowly turn it out 90 degrees to allow the fluid to move into the reservoir. Then turn it one complete turn to thoroughly drain the fluid. ☐ Task completed
7. Retighten the bleeder screw, then collect and discard the fluid. ☐ Task completed
8. Reinstall the cap. ☐ Task completed

Instructor's Comments

____

____

____

BRAKES JOB SHEET 34

Bleeding an ABS Hydraulic System

Name ______________________ Station ______________ Date ______________

NATEF Correlation

This Job Sheet addresses the following NATEF task:

G.5. Bleed the anti-lock brake system's (ABS) front and rear hydraulic circuits.

Objective

Upon completion of this job sheet, you will be able to bleed the anti-lock brake system's front and rear hydraulic circuits.

Tools and Materials

Basic hand tools

Protective Clothing

Goggles or safety glasses with side shields

Describe the vehicle being worked on:

Year ________ Make ____________________ Model ____________________

VIN ____________________ Engine type and size ____________________

PROCEDURE

NOTE: *The front brakes of most antilock brake systems can be bled in the conventional manner, with or without the accumulator being charged. However, bleeding the rear brakes requires a full-charged accumulator or a pressure bleeder attached to the reservoir cap opening, with a minimum of 35 psi.*

Bleeding the ABS Circuits

NOTE: *Bleeding the ABS hydraulic circuits must be performed any time the braking system is opened just as in the base brake system. This can be done in one of two ways depending on the recommendation of the vehicle's manufacturer. The first method requires the use of a diagnostic scan tool or other electronic breakout box bleeder adapter that can trigger pump operation. The pump purges the ABS lines of air. The air is vented out through the brake fluid reservoir. Once the ABS circuits are cleared of air, the base brake system can then be pressure or manually bled. Not all ABS systems require a special scan tool or breakout box for system bleeding. Some systems are equipped with bleeder valves or recommend cracking open brake tube connections at certain points in the system while using pressure or manual bleeding methods similar to those used on the base brake system. A typical ABS bleeding procedure with an ABS tester follows:*

1. Place the vehicle on a level surface with the wheels blocked. Put the transmission in neutral for manual transmission and in park for automatics. ☐ Task completed

2. Release the parking brake. ☐ Task completed
3. Disconnect the electronic system at the connection specified in the service manual and install the tester or bleeder adapter. ☐ Task completed
4. Fill the brake fluid reservoir that serves both the master cylinder and hydraulic module to the maximum fill line. ☐ Task completed
5. Start the engine and allow it to idle for several minutes. Some systems now require that you bleed off line pressure from a service fitting using a special bleeder adapter. If this is required, refill the reservoir to replenish lost fluid after you have done this. ☐ Task completed
6. Turn the selector on the tester to the proper mode for bleeding the system. ☐ Task completed
7. Firmly depress the brake pedal and activate the tester. You should feel a kickback on the brake pedal. Cycle the pump for the recommended time in all recommended test modes. ☐ Task completed
8. After bleeding the ABS system, refill the reservoir. ☐ Task completed
9. Bleed the base brake system. ☐ Task completed

System Bleeding with Accumulator Charged

1. Once accumulator pressure is available to the system, the rear brakes can be bled by opening the rear brake caliper bleeder screws, one at a time, for 10 seconds while holding the brake pedal in the applied position with the ignition switch in the run position. ☐ Task completed
2. Repeat this process until an air-free flow of brake fluid has been observed at each wheel, then close the bleeder screws. ☐ Task completed
3. Pump the brake pedal several times to complete the bleeding procedure. ☐ Task completed
4. Adjust the brake fluid level in the reservoir to the maximum level with a fully charged accumulator. ☐ Task completed

System Bleeding with Pressure Bleeder

1. Attach the pressure bleeder to the reservoir cap opening and maintain a minimum of 35 psi on the system. ☐ Task completed
2. With the brake pedal at rest and the ignition switch off, open the rear wheel bleeder screws, one at a time, for 10 seconds. ☐ Task completed
3. Once an air-free flow of brake fluid has been observed at each wheel, close the bleeder screws and place the ignition switch in the run position. ☐ Task completed
4. Pump the brake pedal several times to complete the bleeder procedure. ☐ Task completed
5. Siphon off the excess fluid in the reservoir to adjust the level to the maximum level with a full-charged accumulator. ☐ Task completed
6. Describe any difficulties you had while completing this job sheet.

____

____

____

____

Instructor's Comments

BRAKES JOB SHEET 35

Servicing ABS Components

Name ______________________ Station ________________ Date ________________

NATEF Correlation

This Job Sheet addresses the following NATEF task:

G.6. Remove and install anti-lock brake system (ABS) electrical/electronic and hydraulic components.

Objective

Upon completion of this job sheet, you will be able to remove and install anti-lock brake system (ABS) electrical/electronic and hydraulic components.

Tools and Materials

Electronic component locator	Hand tools
Syringe	Lift
Service manual	Service manual

Protective Clothing

Goggles or safety glasses with side shields

Describe the vehicle being worked on:

Year __________ Make ________________________ Model ________________________

VIN ______________________ Engine type and size __________________________

Type of ABS ____

PROCEDURE

NOTE: *Every manufacturer and every design of ABS has a different procedure for component removal and installation. Make sure you follow the procedure given in the service manual.*

Hydraulic Components

1. Relieve the accumulator pressure. How did you do that?

   ____

   ____

2. Disconnect the electrical connector from the pressure switch and the motor. ☐ Task completed
3. Use a clean syringe to remove about half of the fluid in the master cylinder reservoir. ☐ Task completed
4. Unscrew the accumulator from the hydraulic module. Then remove the O-ring from the accumulator. ☐ Task completed
5. Disconnect the high-pressure hose from the pump. ☐ Task completed

6. Disconnect the wire retaining clip. Then pull the return hose out of the pump body. ☐ Task completed
7. Remove the bolt that retains the pump and motor to the hydraulic module. ☐ Task completed
8. Remove the pump and motor assembly by sliding it off the locating pin. ☐ Task completed
9. Install the new pump and motor assembly in the reverse of the procedure for removing it. ☐ Task completed
10. Other parts are removed and installed in the same way as conventional brake system parts. What other parts do you need to replace and what is the procedure for doing so?

____

____

____

____

Problems Encountered

____

____

____

Electrical/Electronic Components

1. Before removing or installing any electronic component, know how you will prevent extreme electronic discharge that may damage the components. What will you do?

____

____

2. Disconnect the negative terminal of the battery. ☐ Task completed
3. Disconnect the electrical connector to the component. ☐ Task completed
4. Loosen and remove the mounting bolts or screws for the component. ☐ Task completed
5. Remove the part. ☐ Task completed
6. Install the new part in the reverse order of the procedure for removing it. ☐ Task completed

Problems Encountered

____

____

____

Instructor's Comments

____

____

____

BRAKES JOB SHEET 36

Test a Wheel Speed Sensor and Adjust Its Gap

Name ______________________ Station ______________ Date ______________

NATEF Correlation

This Job Sheet addresses the following NATEF task:

G.7. Test, diagnose, and service ABS speed sensors, toothed ring (tone wheel), and circuits using a graphing voltmeter (GMM), or digital storage oscilloscope (DSO) (includes output signal, resistance, shorts to voltage/ground, and frequency data).

Objective

Upon completion of this job sheet, you will be able to service, test, and adjust anti-lock brake system (ABS) speed sensors.

Tools and Equipment

Lab scope

DMM

Feeler gauge

Service manual

Protective Clothing

Goggles or safety glasses with side shields

Describe the vehicle being worked on:

Year ________ Make ____________________ Model ____________________

VIN ____________________ Engine type and size ____________________

Describe the type of ABS found on the vehicle:

____

Describe the lab scope and DMM that will be used:

____

PROCEDURE

1. Raise the vehicle on a frame contact hoist. Make sure the wheel that has the sensor that will be tested is free to rotate. ☐ Task completed

2. Turn the ignition switch to the run position. ☐ Task completed

3. Visually inspect the wheel speed sensor pulsers for chipped or damaged teeth. Record your findings:

____

____

____

4. Check the wheel sensor's wiring harness for any damage. Record your findings:

____

____

____

5. Connect the lab scope across the wheel sensor. ☐ Task completed

6. Spin the wheel, by hand, and observe the waveform on the scope. You should notice that as the wheel begins to spin, the waveform of the sensor's output should begin to oscillate above and below zero volts. The oscillations should get taller as speed increases. If the wheel's speed is kept constant, the waveform should also stay constant. Record what you observed:

____

____

____

7. Turn off the ignition. ☐ Task completed

8. Locate the air gap specifications for the sensor. The specified gap is: ☐ Task completed

____

9. Use a feeler gauge to measure the air gap between the sensor and its pulser (rotor) all the way around while rotating the drive shaft, wheel, or rear hub bearing unit by hand. Record the measured gap: ________________ ☐ Task completed

10. Does the gap vary while the wheel is rotated? If so, what does this suggest?

____

____

____

11. If the gap is not within specifications, what needs to be done?

____

____

____

12. How would an incorrect air gap at the wheel sensor affect ABS operation?

____

____

____

Problems Encountered

Instructor's Comments

BRAKES JOB SHEET 37

Checking the Vehicle When There Are ABS Concerns

Name ______________________ Station ______________ Date ______________

NATEF Correlation

This Job Sheet addresses the following NATEF task:

G.8. Diagnose anti-lock brake system (ABS) braking concerns caused by vehicle modifications (tire size, curb height, final drive ratio, etc.).

Objective

Upon completion of this job sheet, you will be able to diagnose anti-lock brake system braking concerns caused by vehicle modifications.

Tools and Materials

Tire pressure gauge	Chalk
Machinist's rule or tape measure	Service manual

Protective Clothing

Goggles or safety glasses with side shields

Describe the vehicle being worked on:

Year ________ Make ____________________ Model ____________________

VIN __________________ Engine type and size ____________________

PROCEDURE

NOTE: *Modifications to a vehicle with ABS can cause erratic or poor ABS operation. It is important that you know how the vehicle was equipped when it was manufactured. The VIN will help you, as will the various placards on the vehicle.*

Curb Height

1. Check the trunk for extra weight. What is in the trunk? Should it be removed?

____

____

2. What is the recommended tire air pressure?

____

3. Check the tires for normal inflation pressure. Correct the air pressure if necessary. What were your findings?

____

4. Park the car on a level shop floor or alignment rack. ☐ Task completed

5. Find the vehicle manufacturer's specified curb riding height measurement locations in the service manual. Record the specifications here:

____

____

6. Measure and record the right front curb riding height. Your measurement was: ____________

7. Measure and record the left front curb riding height. Your measurement was: ____________

8. Measure and record the right rear curb riding height. Your measurement was: ____________

9. Measure and record the left rear curb riding height. Your measurement was: ____________

10. Compare the measurement results to the specified curb riding height in the service manual. What do you conclude?

____

____

____

____

11. How will incorrect curb riding height affect the operation of ABS?

____

____

____

____

Tire Size

1. Check all of the tires and record their size; include the width, aspect ratio, and diameter. Record your findings.

____

____

2. What size tire was the vehicle originally equipped with?

____

____

3. If there is a size discrepancy, how will this affect ABS operation?

____

____

4. What service do you recommend?

____

____

Final Drive Ratio

1. What final drive ratio was the vehicle originally fitted with?

____

____

2. How do you know this?

____

____

3. Raise the vehicle so you can rotate the drive shaft and the drive wheels are free to rotate. ☐ Task completed

   Make a reference mark on the drive shaft and the rear axle flange. ☐ Task completed

4. Put a reference mark on the inside of the tire or wheel attached to the drive axle and to some visible stationary part of the chassis. ☐ Task completed

5. Rotate the drive shaft and count the number of turns it takes for the wheel to make one complete revolution. Your count represents the final drive ratio. What was it?

____

____

6. Compare this ratio to the specs and explain how a difference would affect ABS operation.

____

____

Instructor's Comments

____

____

____

NOTICE: SOME PARTS OF THIS COPY ARE DIFFERENT THAN THE PREVIOUS PRINTING OF THIS BOOK.

The contents of this book have been updated in response to the recent changes made by NATEF. Most of these changes were minor and involved the rewording of task statements. There were, however, some new tasks added to their list. The additions resulted in the renumbering of the tasks, those changes have also been made to this workbook.

The NATEF task list included here shows where the additions, deletions, and changes were made. There is also a table that shows which Job Sheet correlates to the new NATEF task numbers, as well as the old.

Job Sheets that relate to the new tasks follow the job sheet correlation chart.

NATEF TASK LIST FOR BRAKE SYSTEMS

Legend: everything that is **new** is underlined
everything that has been **deleted** is ~~struck through~~

A.	**General Brake Systems Diagnosis**	
A.1.	Identify and interpret brake system concern; determine necessary action.	Priority Rating 1
A.2.	Research applicable vehicle and service information, such as brake system operation, vehicle service history, service precautions, and technical service bulletins.	Priority Rating 1
A.3.	Locate and interpret vehicle and major component identification numbers (VIN, vehicle certification labels, calibration labels).	Priority Rating 1
~~A~~B.	**Hydraulic System Diagnosis and Repair**	
B.1.	Diagnose pressure concerns in the brake system using hydraulic principles (Pascal's Law).	
~~A.1.~~B.2.	Measure brake ~~and adjust~~ pedal height; determine necessary action.	Priority Rating 2
~~A.2.~~B.3.	Check master cylinder for internal and external leaks and proper operation; determine necessary action.	Priority Rating 2
~~A.3.~~B.4.	Remove, bench bleed, and reinstall master cylinder.	Priority Rating 1
~~A.4.~~B.5.	Diagnose poor stopping, pulling or dragging concerns caused by ~~problems~~ malfunctions in the hydraulic system; determine necessary action.	Priority Rating 1
~~A.5.~~B.6.	Inspect brake lines, flexible hoses, and fittings for leaks, dents, kinks, rust, cracks, bulging or wear; tighten loose fittings and supports; determine necessary action.	Priority Rating 2
~~A.6.~~B.7.	Fabricate and/or install brake lines (double flare and ISO types); replace hoses, fittings, and supports as needed.	Priority Rating 2
~~A.7.~~B.8.	Select, handle, store, and fill ~~install~~ brake fluids to proper level.	Priority Rating 1
~~A.8.~~B.9.	Inspect, test, and/or replace metering (hold-off), proportioning (balance), pressure differential, and combination valves.	Priority Rating 3
~~A.9.~~B.10.	Inspect, test, replace and adjust height (load) sensing proportioning valve.	Priority Rating 3

~~A.10~~B.11.	Inspect, test, and/or replace components of brake warning light system.	Priority Rating 3
~~A.11.~~B.12.	Bleed (manual, pressure, vacuum or surge) brake system.	Priority Rating 1
~~A.12.~~B.13.	Flush hydraulic system.	Priority Rating 3
~~B~~C.	**Drum Brake Diagnosis and Repair**	
~~B~~C.1.	Diagnose poor stopping, noise, pulling, grabbing, dragging or pedal pulsation concerns; determine necessary action.	Priority Rating 1
~~B~~C.2.	Remove, clean (using proper safety procedures), inspect, and measure brake drums; ~~service or replace as needed~~ determine necessary action.	Priority Rating 1
~~B~~C.3.	~~Mount brake drum on lathe, machine braking surface~~ Refinish brake drum.	Priority Rating 2
~~B~~C.4.	Remove, clean, and inspect brake shoes, springs, pins, clips, levers, adjusters/self-adjusters, other related brake hardware, and backing support plates; lubricate and reassemble.	Priority Rating 2
~~B~~C.5.	Remove, inspect, and install wheel cylinders.	Priority Rating 2
~~B~~C.6.	Pre-adjust brake shoes and parking brake before installing brake drums or drum/hub assemblies and wheel bearings.	Priority Rating 1
~~B~~C.7.	Install wheel, torque lug nuts, and make final checks and adjustments.	Priority Rating 1
~~C~~D.	**Disc Brake Diagnosis and Repair**	
~~C~~D.1.	Diagnose poor stopping, noise, pulling, grabbing, dragging, or pedal pulsation concerns; determine necessary action.	Priority Rating 1
~~C~~D.2.	Remove caliper assembly from mountings; clean and inspect for leaks and damage to caliper housing; determine necessary action.	Priority Rating 1
~~C~~D.3.	Clean and inspect caliper mounting and slides for wear and damage; determine necessary action.	Priority Rating 1
~~C~~D.4.	Remove, clean, and inspect pads and retaining hardware; determine necessary action.	Priority Rating 1
~~C~~D.5.	Disassemble and clean caliper assembly; inspect parts for wear, rust, scoring, and damage; replace seal, boot, and damaged or worn parts.	Priority Rating 1
~~C~~D.6.	Reassemble, lubricate, and reinstall caliper, pads, and related hardware; seat pads, and inspect for leaks.	Priority Rating 1
~~C~~D.7.	Clean, inspect, and measure rotor with a dial indicator and a micrometer; follow manufacturer's recommendations in determining need to machine or replace.	Priority Rating 1
~~C.11.~~D.8.	Remove and ~~replace~~ reinstall rotor.	Priority Rating 2
~~C.8~~D.9.	Refinish rotor according to manufacturer's recommendations.	Priority Rating 1
~~C.9~~D.10.	Adjust calipers with integrated parking brake system.	Priority Rating 3
~~C.10~~D.11.	Install wheel, torque lug nuts, and make final checks and adjustments.	Priority Rating 1
~~D~~E.	**Power Assist Units Diagnosis and Repair**	
~~D~~E.1.	Test pedal free travel with and without engine running; check power assist operation.	Priority Rating 2
~~D~~E.2.	Check vacuum supply (manifold or auxiliary pump) to vacuum-type power booster.	Priority Rating 2
~~D~~E.3.	Inspect the vacuum-type power booster unit for vacuum leaks; inspect the check valve for proper operation; determine necessary action.	Priority Rating 2

~~D~~E.4.	Inspect and test hydro-boost system and accumulator for leaks and proper operation; determine necessary action.	Priority Rating 3
~~E~~F.	**Miscellaneous (Wheel bearings, Parking Brakes, Electrical, Etc.) Diagnosis and Repair**	
~~E~~F.1.	Diagnose wheel bearing noises, wheel shimmy, and vibration concerns; determine necessary action.	Priority Rating 1
~~E~~F.2.	Remove, clean, inspect, repack, and install wheel bearings and replace seals; install hub and adjust wheel bearings.	Priority Rating 1
~~E~~F.3.	Check parking brake cables and components for wear, rusting, binding, and corrosion; clean, lubricate, and replace as needed.	Priority Rating 2
~~E~~F.4.	Check parking brake operation; ~~adjust as needed~~ determine necessary action.	Priority Rating 1
~~E~~F.5.	Check operation of parking brake indicator light system.	Priority Rating 3
~~E~~F.6.	Check operation of brake stop light system; ~~adjust and service as needed~~ determine necessary action.	Priority Rating 1
~~E~~F.7.	Replace wheel bearing and race.	Priority Rating 1
F.8.	Inspect and replace wheel studs.	Priority Rating 1
F.9.	Remove and reinstall sealed wheel bearing assembly.	Priority Rating 1
~~F~~G.	**Anti-lock Brake System**	
~~F~~G.1.	Identify and inspect ~~and test~~ anti-lock brake system (ABS) components; determine necessary action.	Priority Rating 2
~~F~~G.2.	Diagnose poor stopping, wheel lock-up, abnormal pedal feel or pulsation, and noise concerns caused by the anti-lock brake system (ABS); determine necessary action.	Priority Rating 2
~~F~~G.3.	Diagnose anti-lock brake system (ABS) electronic control(s) and components using self-diagnosis and/or recommended test equipment; determine necessary action.	Priority Rating 1
~~F~~G.4.	Depressurize high-pressure components of the anti-lock brake system (ABS).	Priority Rating 2
~~F~~G.5.	Bleed the anti-lock brake system's (ABS) front and rear hydraulic circuits.	Priority Rating 2
~~F~~G.6.	Remove and install anti-lock brake system (ABS) electrical/ electronic and hydraulic components.	Priority Rating 3
~~F~~G.7.	~~Service~~, Test, ~~and adjust~~ diagnose and service ABS speed sensors, toothed ring (tone wheel), and circuits using a graphing voltmeter (GMM), or digital storage oscilloscope (DSO) (includes output signal, resistance, shorts to voltage/ground, and frequency data).	Priority Rating 2
~~F~~G.8.	Diagnose anti-lock brake system (ABS) braking concerns caused by vehicle modifications (tire size, curb height, final drive ratio, etc.).	Priority Rating 3
G.9.	Identify traction control system components.	Priority Rating 1

New Task #	Old Task #	Job Sheet #
A.1	NEW	38
A.2	NEW	39
A.3	NEW	39
B.1	NEW	40
B.2	A.1	1
B.3	A.2	2
B.4	A.3	3
B.5	A.4	4
B.6	A.5	5
B.7	A.6	5
B.8	A.7	6
B.9	A.8	7
B.10	A.9	8
B.11	A.10	9
B.12	A.11	10
B.13	A.12	6
C.1	B.1	11
C.2	B.2	12
C.3	B.3	13
C.4	B.4	12
C.5	B.5	14
C.6	B.6	12
C.7	B.7	15
D.1	C.1	16
D.2	C.2	17
D.3	C.3	17
D.4	C.4	17
D.5	C.5	18
D.6	C.6	18
D.7	C.7	19
D.8	C.11	23
D.9	C.8	20
D.10	C.9	21
D.11	C.10	22
E.1	D.1	24
E.2	D.2	25
E.3	D.3	25
E.4	D.4	26
F.1	E.1	27
F.2	E.2	28
F.3	E.3	29
F.4	E.4	30
F.5	E.5	9
F.6	E.6	9
F.7	E.7	28
F.8	NEW	41
F.9	NEW	42
G.1	F.1	31
G.2	F.2	31
G.3	F.3	32
G.4	F.4	33
G.5	F.5	34
G.6	F.6	35
G.7	F.7	36
G.8	F.8	37
G.9	NEW	43

BRAKES JOB SHEET 38

Identifying Problems and Concerns

Name ______________________________ Station __________________ Date __________________

NATEF Correlation

This Job Sheet addresses the following NATEF task:

A.1. Identify and interpret brake system concern; determine necessary action.

Objective

Upon completion of this job sheet, you will be able to define brake system problems or concerns, prior to diagnosing or testing the systems.

Protective Clothing

Goggles or safety glasses with side shields

Describe the vehicle being worked on:

Year ___________ Make __________________________ Model __________________________

VIN _______________________ Engine type and size ___________________________

PROCEDURE

1. Start the engine and describe how the engine seems to be running:

____

____

2. Take the vehicle for a safe road test and pay strict attention to how the brakes operate in all conditions (partial application, moderate application, and hard application). Describe your results here.

____

____

____

____

3. Were there any unusual noises when the brakes were applied? If so, describe them and when they occurred.

____

____

4. Describe how the pedal felt when it was applied.

____

____

5. Did the vehicle tend to pull to one side when the brakes were applied? Explain.

____

____

6. Did the brakes seem to drag? Explain.

____

____

7. Turn off the engine and pump the brakes several times. What happened?

____

____

8. Summarize your findings in detail; include both improper and proper system operation.

____

____

____

____

____

____

____

9. Based on the above, what are your suspicions and conclusions?

____

____

____

____

Instructor's Comments

____

____

____

BRAKES JOB SHEET 39

Gathering Vehicle Information

Name ______________________________ Station __________________ Date __________________

NATEF Correlation

This Job Sheet addresses the following NATEF tasks:

A.2. Research applicable vehicle and service information, such as brake system operation, vehicle service history, service precautions, and technical service bulletins.

A.3. Locate and interpret vehicle and major component identification numbers (VIN, vehicle certification labels, calibration labels).

Objective

Upon completion of this job sheet, you will be able to gather service information about a vehicle and its brake system.

Tools and Materials

Appropriate service manuals

Computer

Protective Clothing

Goggles or safety glasses with side shields

Describe the vehicle being worked on:

Year __________ Make __________________________ Model __________________________

VIN _______________________

PROCEDURE

1. Using the service manual or other information source, describe what each letter and number in the VIN for this vehicle represents.

   ____

   ____

2. Locate the Vehicle Emissions Control Information (VECI) label and describe where you found it.

   ____

3. Summarize what information you found on the VECI label.

   ____

   ____

4. Using a service manual or electronic database, locate the information about the vehicle's brake system. List the major components of the system and describe how the system's pressure is controlled.

____

____

____

____

5. Using a service manual or electronic database, locate and record all service precautions regarding brake system noted by the manufacturer.

____

____

____

____

6. Using the information that is available, locate and record the vehicle's service history.

____

____

____

____

7. Using the information sources that are available, summarize all Technical Service Bulletins for this vehicle that relate to the brake system.

____

____

____

____

Instructor's Comments

____

____

____

BRAKES JOB SHEET 40

Diagnosing Pressure Problems

Name ______________________________ Station __________________ Date __________________

NATEF Correlation

This Job Sheet addresses the following NATEF task:

B.1. Diagnose pressure concerns in the brake system using hydraulic principles (Pascal's Law).

Objective

Upon completion of this job sheet, you will be able to correctly diagnose pressure problems in the brake system.

Tools and Materials

Steel or machinist rule

Service manual

Protective Clothing

Goggles or safety glasses with side shields

Describe the vehicle being worked on:

Year __________ Make __________________________ Model __________________________

VIN ______________________ Engine type and size __________________________

PROCEDURE

1. Visually inspect the entire hydraulic system for signs of leakage and/or damage. Record your results.

   ____

   ____

2. Pump the brake pedal with the engine off to exhaust vacuum in the booster. ☐ Task completed
3. Place a ruler against the car floor, inline with the arc of pedal travel. ☐ Task completed
4. Press the pedal by hand and measure the amount of travel before looseness in the linkage, or freeplay, is taken up. ☐ Task completed
5. Measure at the top or bottom of the pedal, whichever provides the most accurate view. Compare your measurement with the vehicle specifications. Summarize your results.

   ____

   ____

6. Pump the pedal several times and remeasure the travel of the pedal. What were your findings?

____

____

____

____

7. What do the results from step #5 indicate? Explain.

____

____

____

____

8. Use Pascal's Law to explain why you felt what you did at the brake pedal.

____

____

____

____

Instructor's Comments

____

____

____

BRAKES JOB SHEET 41

Inspecting and Replacing Wheel Studs

Name ______________________ Station ______________ Date ______________

NATEF Correlation

This Job Sheet addresses the following NATEF task:

F.8. Inspect and replace wheel studs.

Objective

Upon completion of this job sheet, you will be able to inspect and replace wheel studs in axle flanges, brake rotors, and drums.

Tools and Materials

Bushing tool

Hammer

Punch set

Thread chaser

Protective Clothing

Goggles or safety glasses with side shields

Describe the vehicle being worked on:

Year ________ Make ____________________ Model ____________________

VIN ________________ Tire size ____________ Wheel Diameter ________

PROCEDURE

1. Determine the proper lift points for the vehicle. Describe them here:

   ____

   ____

2. Raise the vehicle. Make sure it is high enough to remove the wheel assembly. Once it is in position, place safety stands under the vehicle or set the mechanical lock on the lift. ☐ Task completed

3. Remove the lug nuts and the tire and wheel assembly. ☐ Task completed

4. Are any wheel studs broken or missing? Explain.

   ____

   ____

5. Carefully inspect the condition of the wheel stud and lug nut threads. Describe their condition.

____

____

6. If the lug nuts are severely damaged, both the lug nut and the associated wheel stud should be replaced. If the lug nut threads are slightly damaged, they can be cleaned and corrected with a thread chaser. What size chaser would you use in the lug nuts of this vehicle?

____

7. If a wheel stud needs to be replaced, the procedure to do so varies with the component the stud in installed in. The studs can be pressed into the axle flange, rotor, or drum. Where are the studs installed on this vehicle?

____

____

8. If the wheel studs are installed in the axle flange they can typically be loosened by tapping the outer end of the stud with a hammer and punch. Attempt to do this and describe what happened.

____

____

9. If the stud did not loosen with the tapping, use a bushing press to press the stud out of the flange. ☐ Task completed

10. Removing a stud from a drum or rotor can be more difficult, depending on how the stud is retained. Carefully study the drums and/or rotors and describe how the studs are retained.

____

____

11. If the stud is held by swaging, the raised metal around the base of the stud must be removed with a special cutting tool. ☐ Task completed

12. After the retaining metal has been removed, the stud can be tapped or pressed. Make sure you use the correct tools and don't damage or distort the drum or rotor. Describe any difficulties you had in doing this.

____

____

13. Before installing a new stud, make sure it is the same diameter and length as the original one and has the same length of serrations. ☐ Task completed

14. Install the new stud into its bore. ☐ Task completed

15. Place several flat washers over the stud threads. ☐ Task completed

16. Install and tighten a lug nut, with its flat side against the flat washers, onto the stud. The stud will be pulled into position as the lug nut is tightened. ☐ Task completed

17. Loosen the lug nut and remove the washers. Check to make sure the stud is fully seated. ☐ Task completed

18. Before installing the wheel and tire assembly, make sure the tire is inflated to the recommended inflation. What is that inflation?

____

19. Install the wheel assembly onto the vehicle and start each of the lug nuts. ☐ Task completed

20. Hand tighten each lug nut so that the wheel is fully seated against the hub. Then tighten the lug nuts to specifications and in a staggered or star pattern. The order you used to tighten the lug nuts was:

____

The recommended torque spec is: ____________________

Instructor's Comments

____

____

____

BRAKES JOB SHEET 42

Servicing Sealed Wheel Bearings

Name ______________________ Station ______________ Date ______________

NATEF Correlation

This Job Sheet addresses the following NATEF task:

F.9. Remove and reinstall sealed wheel bearing assembly.

Objective

Upon completion of this job sheet, you will be able to replace sealed front wheel bearings.

Tools and Materials

Lift
Torque wrench
Hydraulic press with fixtures
Various drivers and pullers
Brass drift
Soft-faced hammer
CV joint boot protector

Protective Clothing

Goggles or safety glasses with side shields

Describe the vehicle being worked on:

Year ________ Make ________________ Model ________________

VIN ________________ Engine type and size ____________________

PROCEDURE

NOTE: *This is a typical procedure. Check with the service manual for the exact procedure you should follow and mark the differences in the procedure as you progress through these steps:*

1. Have someone apply the brakes with the vehicle's weight on the tires. Loosen the hub nut and wheel lug nuts. Then remove the axle hub nut. ☐ Task completed
2. Jack up and support the vehicle on safety stands. Remove the tire and wheel assembly. ☐ Task completed
3. Install a boot protector over the CV joint boot. ☐ Task completed
4. Unbolt the brake caliper, move it out of the way, and suspend it with wire. Make sure the flexible brake hose isn't supporting the weight of the caliper. ☐ Task completed
5. Remove the brake rotor. ☐ Task completed

6. Loosen and remove the pinch bolt that holds the lower control arm to the steering knuckle and separate the lower ball joint and tie-rod end from the knuckle. ☐ Task completed

7. On some vehicles, an eccentric washer is located behind one or both of the bolts that connects the spindle to the strut. These eccentric washers are used to adjust camber. Mark the exact location and position of these washers on the strut before you remove them. ☐ Task completed

8. Remove the hub-and-bearing to knuckle retaining bolts. ☐ Task completed

9. Remove the cotter pin and castellated nut from the tie-rod end and pull the tie-rod end from the knuckle. The tie-rod end can usually be freed from its bore in the knuckle by using a small gear puller or by tapping on the metal surrounding the bore in the knuckle. ☐ Task completed

10. Remove the pinch bolts holding the steering knuckle to the strut, and then disconnect the knuckle from the vehicle. ☐ Task completed

11. With a soft-faced hammer, tap on the hub assembly to free it from the knuckle. The brake backing plate may come off with the hub assembly. ☐ Task completed

12. Install the appropriate puller and press the hub-and-bearing assembly from the half shaft. ☐ Task completed

13. Pull the half shaft away from the steering knuckle. Do not allow the CV joint to drop while you remove the hub assembly. Support the half shaft and CV joint from the backside of the knuckle. ☐ Task completed

14. Remove the snap ring or collar that retains the wheel bearing. ☐ Task completed

15. Mount the knuckle on a hydraulic press so that it rests on the base of the press with the hub free. Using a driver slightly smaller than the inside diameter of the bearing, press the bearing out of the hub. On some vehicles, this will cause half of the inner race to break out. Since half of the inner race is still in the hub, it must be removed with a small gear puller. ☐ Task completed

16. Inspect the bearing's bore in the knuckle for burrs, score marks, cracks and other damage and record your findings and recommendations.

    ____

    ____

17. Lightly lubricate the outer surface of the new bearing assembly and the bore in the knuckle. ☐ Task completed

18. Press the new bearing into the knuckle making sure the pressing tool is contacting the outer diameter of the bearing. ☐ Task completed

19. Install the retaining snap rings or collar. ☐ Task completed

20. Install the brake splash shield onto the knuckle. ☐ Task completed

21. Press a new steering knuckle seal into the knuckle and lubricate the lip seal and coat the inside of the knuckle with a thin coat of clean grease. ☐ Task completed

22. Slide the knuckle assembly over the half shaft. ☐ Task completed

23. Mount the knuckle to the strut and tighten the bolts. Make sure the camber marks to the strut are positioned properly. ☐ Task completed

24. Reinstall the lower control arm's ball joint and tie-rod end to the steering knuckle. Then mount the rotor and brake caliper. Torque all bolts to specifications. What are the specifications for those bolts?

____

____

25. Install the tire/wheel assembly and lower the vehicle. ☐ Task completed

26. Tighten the new axle hub nut and wheel lug nuts to specifications. What are the specifications for those nuts?

____

____

27. Stake or use a cotter pin to keep the hub nut in place after it is tightened. ☐ Task completed

28. Road test the vehicle and recheck the torque on the hub nut. Summarize the results of your road test.

____

____

____

____

29. Check the vehicle's wheel alignment. Indexing the eccentric washers during disassembly will only get the alignment close after reassembly. How far off was the alignment after your repair?

____

____

Instructor's Comments

____

____

____

BRAKES JOB SHEET 43

Identifying Traction Control Components

Name ______________________ Station ______________ Date ______________

NATEF Correlation

This Job Sheet addresses the following NATEF task:

G.9. Identify traction control system components.

Objective

Upon completion of this job sheet, you will be able to gather service information about the components of a vehicle's traction control system.

Tools and Materials

Appropriate service manuals

Computer

Protective Clothing

Goggles or safety glasses with side shields

Describe the vehicle being worked on:

Year ________ Make ____________________ Model __________________

VIN ____________________

PROCEDURE

1. What does the manufacturer call the traction control system that the vehicle is equipped with?

   ____

2. Using the service manual or other information source, describe the basic operation of the traction control system used on this vehicle.

   ____

   ____

   ____

   ____

3. List all of the major parts that are part of the traction control system.

   ____

   ____

   ____

   ____

4. Where are these major parts located on the vehicle?

____

____

____

____

5. Which of these parts is used only for traction control?

____

____

____

____

____

Instructor's Comments

____

____

____